Scourge

ALSO BY JONATHAN B. TUCKER:

Ellie: A Child's Fight Against Leukemia
Toxic Terror: Assessing Terrorist Use of Chemical and Biological Weapons (Editor)

Scourge

The Once and Future Threat of Smallpox

Jonathan B. Tucker

Atlantic Monthly Press
New York

Published simultaneously in Canada
Printed in the United States of America

Library of Congress Cataloging-in-Publication Data
Tucker, Jonathan B.
    Scourge : the once and future threat of smallpox / Jonathan Tucker.
      p. cm.
    Includes bibliographical references.
    ISBN 0-87113-830-1
    1. Smallpox—History.   I. Title.
  RC183.1 .T83 2001
  616.9'12'009—dc21   2001022913

Design by Laura Hammond Hough

Atlantic Monthly Press
841 Broadway
New York, NY 10003

01 02 03 04   10 9 8 7 6 5 4 3 2

A portion of the earnings from the publication of *Scourge* has been donated to Seva Foundation, Berkeley, California (visit www.seva.org).

CONTENTS

ACKNOWLEDGMENTS

Most of this book was written during a 1999–2000 Robert Wesson Fellowship in Scientific Philosophy and Public Policy at Stanford University's Hoover Institution on War, Revolution, and Peace. I am deeply grateful to Sidney Drell, Abraham Sofaer, and Thomas Henriksen at Hoover for helping to arrange the fellowship year on my behalf. Thanks are also due to William Potter, director of the Center for Nonproliferation Studies at the Monterey Institute of International Studies, for granting me a one-year leave of absence.

I am indebted to D. A. Henderson, who generously donated many hours of his time to help me with research leads and to correct technical errors. Several other individuals also deserve thanks for reading part or all of the manuscript and providing useful comments: Ken Alibek, Judith Bale, Ken Bernard, Lawrence Brilliant, Peter Carrasco, Ciro de Quadros, Joseph Esposito, Margaret Hamburg, Peter Jahrling, Robert Kadlec, Ali Khan, Stephen Ostroff, Robert Siegel, Tara O'Toole, and Alan Zelicoff.

Finally, I would like to thank Martha Kaplan, my agent, and Joan Bingham, my editor at Atlantic Monthly Press, for their invaluable advice and support.

Jonathan B. Tucker
Washington, D.C.

One

MONSTER ON DEATH ROW

In a maximum-security facility in Atlanta, the world's most dangerous prisoner sits in solitary confinement, awaiting execution. Wanted for the torture and death of millions of people, this mass murderer was captured in a global dragnet lasting more than a decade. Although the prisoner has been condemned to death, the jailers are debating whether or not to carry out the sentence. Some believe that studying the killer's methods could help to develop better defenses against such crimes, yet others fear that the prisoner could escape and wreak mayhem on an even greater scale. While the debate continues, the execution has been postponed.

The world's most dangerous prisoner is the smallpox virus, and it is held inside two padlocked freezers in a secure room at the U.S. Centers for Disease Control and Prevention in Atlanta. Some 450 samples of the virus in neatly labeled, half-inch plastic vials are arrayed on metal racks and immersed in a bath of liquid nitrogen that keeps them deep-frozen at −94 degrees Fahrenheit. Access to the smallpox repository requires two sets of keys controlled by different people; closed-circuit television cameras and electronic alarm systems maintain continuous surveillance. A second set of smallpox virus stocks lies in a similar vault at a Russian laboratory in Siberia.

The scientific name for the smallpox virus is variola, the diminutive of the Latin word *varius* (spotted) or *varus* (pimple). Consisting of little more than a set of genetic instructions in a long ribbon of DNA, coiled up inside a biscuit-shaped protein shell, the variola virus cannot grow or metabolize and has no means of locomotion. Its sole function is self-replication, which it accomplishes by entering human cells and commandeering their biochemical machinery to churn out more virus particles. When variola existed in the wild, it came in two distinct varieties: *Variola major* caused a serious disease that killed between 10 percent and 30 percent of its victims, whereas *variola minor* gave rise to a much milder illness called alastrim, with a case mortality rate of less than 1 percent. Because the two types of smallpox virus produced similar symptoms, it is not known why one was so much more lethal than the other.

Now confined to a few laboratory freezers, variola major once rampaged through the human species and caused the most feared of deadly scourges. After a two-week incubation period, smallpox racked the body with high fever, headache, backache, and nausea, and then peppered the face, trunk, limbs, mouth, and throat with hideous, pus-filled boils. Patients with the infection were in agony—their skin felt as if it was being consumed by fire, and although they were tormented by thirst, lesions in the mouth and throat made it excruciating to swallow. The odor of a smallpox ward was oppressive: The rash gave off a sweetish, pungent smell reminiscent of rotting flesh. For those who survived, the disease ran its course in a few weeks. Pustule formation concluded on days eight to ten of the illness, after which the boils scabbed over and were gradually reabsorbed. On days fifteen to twenty, the crusty dry scabs separated and fell off, leaving depigmented areas of skin that later turned into ugly, pitted scars.

Even as smallpox victims were suffering the torments of the disease, they were spreading it to others. Lesions in the patient's mouth and throat shed millions of virus particles into the saliva and mucus, so that talking or sneezing expelled virus-laden droplets that floated in the air and could be inhaled. The virus was also present in patients' urine and in pus from unhealed skin lesions. When clothing and bed linens contaminated with dried pus were handled, virus particles could be resuspended in the air, so that laundry workers who washed the sheets and blankets of smallpox patients were at great risk of infection. The corpses of smallpox victims were also dangerously contaminated and could spread the disease to undertakers or to family members who prepared a loved one's body for ritual burial.

Over the course of human history, smallpox claimed hundreds of millions of lives, far more than plague—the dreaded Black Death of the Middle Ages—and all the wars of the twentieth century combined. Although those lucky enough to survive a bout with smallpox acquired lifelong immunity, they usually suffered some type of permanent damage. Nearly all were disfigured with pockmarks, and one in ten was rendered partially or completely blind. In addition, smallpox significantly increased the risk of miscarriage in pregnant women.

As recently as 1967, the disease sickened between ten million and fifteen million people each year in forty-three countries and caused an estimated two million deaths. On May 8, 1980, however, the World Health Organization (WHO) declared that humanity had finally been freed from the torments of smallpox, the culmination of a global campaign lasting more than a decade and employing up to 150,000 health workers at various times. The conquest of smallpox, the first—and so far, only—infectious disease to have been eradicated from nature by human effort, was among the greatest medical achievements of the twentieth century.

After the WHO formally certified the eradication of smallpox in May 1980, all member countries agreed to stop vaccinating their civilian populations because the potential risk of complications from the vaccine now outweighed the tiny chance that smallpox might re-emerge from natural sources. Since then, the horrors of the disease have faded from public consciousness like the memory of a nightmare. Fewer and fewer individuals bear the round, mottled scar of a smallpox vaccination on their upper arm or thigh, let alone the disfiguring pockmarks that were once the hallmark of the disease. But although some would relegate the history of smallpox to the dusty shelves of a medical library, such complacency would be premature. In 1992, a senior Russian official defected to the United States and told the CIA that the Soviet Union, even as it had supported the smallpox eradication campaign with vaccine and expertise, had secretly developed the virus into a military weapon and stockpiled enough of it to kill millions of people. News of the Soviet betrayal sparked official concern in Washington, London, and other capitals that samples of the virus might fall into the hands of "rogue" states and terrorist organizations.

Because the immunity induced by the smallpox vaccine fades after about a decade, most of the world is now susceptible to infection. Responding to this potential threat, the United States and other countries are undertaking urgent efforts to strengthen their medical defenses against this supposedly eradicated disease. Every human being on the planet has a stake in the fate of the smallpox virus, for we are all ultimately at risk.

SMALLPOX AND CIVILIZATION

Although smallpox was a uniquely human disease, its causative agent, variola virus, belongs to the genus of orthopoxviruses ("true pox" viruses), whose members include buffalopox, camelpox, cowpox, monkeypox, mousepox, rabbitpox, and raccoonpox. These viruses are generally named after their primary host, but some have turned out to be misnomers: Cowpox and monkeypox are actually carried by rodents and only occasionally infect cows or monkeys. The name "chickenpox" is also misleading: The virus infects only humans and is not a poxvirus at all but a member of the unrelated class of herpesviruses. (The word "chicken" in chickenpox may have been derived from the French word *chiche,* meaning chickpea, referring to the size of the lesion, or from the Old English word *gicans,* meaning itch; "pox" refers to any rash consisting of pustules, or skin lesions filled with pus.)

Variola virus appears to have evolved from an animal poxvirus in Central Africa thousands of years ago, when dense tropical forests blanketed the continent all the way to present-day Egypt. The most likely scenario is that the progenitor of variola virus circulated in a species of wild rodent whose populations were large and bred rapidly. Small bands of prehistoric humans hunted these rodents for food and occasionally became infected with the virus. Although variola caused a mild or symptomless condition in its wild animal host, with whom it had co-evolved over millions of years, it produced a severe illness when it jumped the species barrier to humans.

Individuals infected with variola virus either died or became immune for life, so the disease quickly burned itself out in small, isolated populations. Because acquired immunity was not passed down from parents to their offspring, however, each new generation was vulnerable to infection. Smallpox therefore flared up for a limited period whenever a susceptible band of hunter-gatherers came in contact with infected rodents. Such sporadic outbreaks may have occurred in prehistoric human groups over a protracted period.

Around 9000 B.C., the first agricultural settlements grew up in the fertile river basins of ancient Egypt and Mesopotamia. As humans congregated into towns and cities, their populations became large and dense enough to provide a pool of susceptible individuals through new births and the migration of adults from other areas, enabling smallpox to be transmitted continually without ever running out of hosts. The demographic threshold at which a population could sustain the disease on an ongoing basis was around 200,000 people.

Over the ensuing millennia, genetic variants or "strains" of variola virus emerged through random mutation. Those new strains that could spread most effectively through the population gradually predominated. Eventually, variola virus diverged genetically from its rodent progenitor to the point where it was uniquely specialized for infecting and replicating inside human cells.

The earliest physical evidence for the presence of smallpox in ancient Egypt is a striking rash of yellow pustules on the mummified face and hands of

Ramses V, a pharaoh who died at age forty in 1157 B.C. and whose well-preserved remains are on display in the Cairo Museum. Traders carried smallpox from Egypt to India, where Sanskrit medical texts describe epidemics as early as 1500 B.C. The disease arrived in China by 1122 B.C., apparently imported by the Huns, since the Chinese called it "Hunpox."

Smallpox had a major impact on the history of the ancient world. According to the Greek historian Thucydides, an epidemic suggestive of smallpox struck Athens around 430 B.C., killing a third of the city-state's population and contributing to its defeat by Sparta in the Peloponnesian War. In the fourth century B.C., Alexander the Great's army was ravaged by the disease during a campaign in India. The Roman Emperor Marcus Aurelius died of smallpox in A.D. 180, accelerating the decline of the Roman Empire. During the Elephant War in A.D. 570, Abyssinian troops on elephants besieging the Arab capital of Mecca were decimated by an outbreak of smallpox, an incident described metaphorically in the Koran.

In the seventh and eighth centuries, Arab armies carried smallpox across North Africa and into the Iberian Peninsula. From the eleventh through the thirteenth centuries, crusaders returning to Europe from the Holy Land and traders plying the Silk Road to China dispersed the disease widely. In Great Britain in the late fifteenth century, the pustular skin rash came to be called the "small pockes" (from *pocke,* meaning sac) to distinguish it from syphilis, then known as the "great pockes."

Two factors—one epidemiological, the other genetic—influenced the toll that smallpox and other infectious diseases inflicted on human populations. When an infected foreigner introduced variola virus into an isolated community that had had no prior exposure to it, the virus spread explosively, giving rise to a devastating "virgin soil" epidemic that killed people of all ages. As the victims died off or became immune, the disease waned. Only when enough time had passed for a new cohort of susceptible hosts to accumulate in the population through births and migration—like a fresh growth of trees after a forest fire—could the reintroduction of the virus trigger another major epidemic.

Somewhat paradoxically, the longer a society lived with smallpox, the less severe its demographic impact became. In densely populated urban areas, the disease smoldered continuously at a low level and the intervals between major outbreaks were fairly short. As a result, nearly everyone who survived into adulthood was immune and the victims were mainly small children. Historian William H. McNeill has argued that because infants or toddlers were far more easily replaced than adolescents or adults, the economic and psychological burdens of smallpox on society were much less than they were when the young and old died indiscriminately. In isolated or rural areas, however, where the population was too small or diffuse to support continuous transmission, smallpox was introduced periodically from outside. Since people living in such areas were less likely to encounter the disease in childhood, periodic "boom and bust" epidemics struck down adolescents and young adults as well as infants. Iceland, for example, experienced twenty-one major epidemics of smallpox between the thirteenth and nineteenth centuries, including a massive outbreak in 1707–09 that wiped out 26 percent of the island's population.

A second factor that may have affected the mortality caused by smallpox was the evolution of genetic resistance. People vary in their susceptibility to infectious diseases, and resistant individuals are more likely to survive epidemics. Thus, in populations exposed to smallpox for many generations, one would expect that resistance genes would become more prevalent. When smallpox first appeared in the Old World, it apparently caused very high death rates. As the most susceptible individuals were culled from the population, more resistant survivors came to dominate the gene pool and the impact of the disease gradually became less devastating. Societies that had never been exposed to smallpox, however, retained their genetic vulnerability.

Because of the twin processes of epidemiological adjustment and the evolution of genetic resistance, countries with populations large enough to sustain and adapt to communicable diseases such as smallpox, measles, and mumps acquired a powerful advantage over smaller, more isolated societies that did not experience these infections on a continuous basis. The European explorers and colonists who crossed the oceans to the newly discovered continents of America, Australia, and southern Africa were nearly all immune to smallpox, having survived it in childhood, but they introduced the virus to previously isolated populations that were highly vulnerable.

The Spanish conquest of Mexico and Peru provides a dramatic example. In 1507, Spanish colonists occupied the Caribbean islands of Hispaniola (today, the Dominican Republic and Haiti) and Cuba, and subjected the natives to inhuman working conditions in mines and plantations. The resulting high death rates depleted the indigenous labor pool, leading the Spanish to import slaves from West Africa as replacements. Since many African slaves came from regions where smallpox was endemic, the slave trade introduced the disease to the Americas. Smallpox began its rampage through the New World with an epidemic on Hispaniola in 1518 that killed up to half the native population. The following year, the disease spread to Cuba, and from there it moved across the Antilles to the Mexican mainland.

In 1519, the Spanish conquistador Hernando Cortes, in quest of slaves and gold, led an army of five hundred men, sixteen horses, and several pieces of artillery to explore and conquer Mexico, which was then ruled by the Aztec civilization with a population of many millions. Cortes managed to persuade several local Amerindian tribes to become his allies by promising to free them from Aztec domination. Thus reinforced, the Spanish troops entered the Aztec capital of Tenochtitlan (now Mexico City) without bloodshed and took the Aztec emperor, Montezuma, hostage. After occupying Tenochtitlan for six months, Cortes marched off to confront a rival group of conquistadors led by Panfilo de Narvaez, who had been sent by the Spanish authorities in Cuba to arrest Cortes for disobedience and treason. Cortes left the Aztec capital in the hands of a repressive lieutenant, who proceeded to slaughter six hundred unarmed native worshipers at a religious ceremony. This atrocity triggered an uprising led by Montezuma's brother Cuitlahuac, and the numerically superior Aztec army inflicted heavy casualties on the Spanish. After defeating de Narvaez, Cortes rushed back to Tenochtitlan with reinforcements in late June 1520, but the Aztecs beat his troops decisively seven days later. Having lost nearly a third of his men, Cortes retreated to a coastal settlement.

Had the Aztecs continued their pursuit, they could have expelled the defeated Spanish from Mexico. They failed to do so, however, because smallpox had broken out in Tenochtitlan, spread by an African slave who had accompanied de Narvaez's forces. The virgin-soil epidemic scythed through the native population, killing nearly half of the young and old alike. Among the dead were Cuitlahuac and other senior Aztec leaders and warriors. The epidemic gave Cortes and his men time to regroup, and they laid siege to Tenochtitlan for seventy-five days in the summer of 1521. This time, the demoralized defenders put up relatively little resistance, and the city fell on August 13. When the Spanish troops entered the Aztec capital, it was strewn with the bodies of smallpox victims. "I have read about the destruction of Jerusalem, but I do not think the mortality was greater there than in Mexico," wrote Cortes's chronicler Bernal Diaz. "Indeed, the stench was so bad that no one could endure it . . . and even Cortes was ill from the odors which assailed his nostrils."

Because smallpox had decimated the Amerindian population while leaving the Spanish invaders unscathed, it had a shattering psychological impact. According to historian McNeill, the Aztecs interpreted the selective pestilence as a demonstration of the superior power of the God the Spanish worshiped. As a result, Cortes and his ragtag army were able to sub-jugate the Aztec empire of some 12.5 million people. Within a generation, the culture, religion, and language of the Spanish colonizers had displaced local traditions, institutions, and beliefs.

From Mexico, smallpox spread to Guatemala and continued southward, reaching the Incan lands (present-day Peru) around 1525. There the dis-ease wreaked similar havoc, killing the emperor Huayna Capac and most of his family, including the son he had designated as his heir, and other se-nior military commanders. An ensuing succession struggle divided the Incan empire into two factions that waged a bloody civil war. By the time Francisco Pizarro arrived in Cuzco in 1532 with 67 horsemen and 110 footsoldiers, they met no serious military resistance. In 1563, Portuguese colonizers brought smallpox to Brazil, where it wiped out entire indigenous tribes.

Importations of smallpox also decimated the native peoples of North America, facilitating the European colonization of the continent. In 1616–19, a smallpox epidemic cut down almost nine-tenths of the Indian popu-lation in the Massachusetts Bay area shortly before the arrival of the Pilgrim settlers in 1620. Pilgrim leader John Winthrop attributed the epidemic to an act of God. "The natives, they are near all dead of the smallpox," he wrote, "so as the Lord hath cleared our title to what we possess."

In 1634, William Bradford, the governor of the Plymouth Colony, viv-idly described an outbreak of smallpox among the Connecticut Indians. The victims, he reported, "fall into a lamentable condition, as they lye on their hard matts, ye poxe breaking and mattering, and running one into another, their skin cleaving (by reason thereof) to the matts they lye on; when they turn them, a whole side will flea off at once . . . and they will be all of a gore blood, most fearful to behold; and then being very sore, what with could and other distempers, they dye like rotten sheep." Ac-cording to one estimate, the native population of North and South

America, which had numbered roughly 72 million when Columbus landed in 1492, had been diminished by warfare and disease to about 600,000 in 1800, a staggering decline.

Smallpox was a democratic scourge, afflicting people of every race, class, and social position: Cosmetics were invented to mask the ruined complexions of the rich. The disease killed royalty as well as commoners, disrupting dynasties and alliances and repeatedly changing the course of world history. Rulers of Japan, Siam, Ceylon, Ethiopia, and Burma died of smallpox, and several European crowned heads met the same fate, including Queen Mary II of England in 1694, King Louis I of Spain in 1724, Tsar Peter II of Russia in 1730, Queen Ulrika Eleanora of Sweden in 1741, and King Louis XV of France in 1774. In Austria between 1654 and 1767, the deaths from smallpox of eleven members of the reigning Hapsburg dynasty caused four shifts in the line of succession in as many generations.

Ironically, the ubiquity of smallpox in seventeenth-century England had the beneficial effect of fostering the establishment of higher education in the American colonies. The more prosperous colonists had routinely sent their male offspring to be educated at Oxford or Cambridge, but there they ran the risk of contracting smallpox, which, by virtue of their more isolated upbringing, they had previously escaped. Seeking to avoid this risk, the colonists decided to educate their sons at home by founding new institutions of higher learning, including Harvard College in 1636, the College of William and Mary in 1693, and Yale College in 1701.

By the turn of the eighteenth century, smallpox had replaced bubonic plague, which was more episodic, as Europe's most devastating and feared disease. London experienced five major smallpox epidemics between 1719 and 1746; outbreaks also occurred in Rome, Berlin, Geneva, and other European cities. Smallpox was carried by ship from India to South Africa in 1713, and it reached Australia shortly after the founding of the European settlement there in 1789. By the end of the eighteenth century, smallpox was killing some 400,000 Europeans a year, mostly children and young adults, and was responsible for a third of all cases of blindness.

The disease cast a long shadow over daily life. As British historian Thomas Babington Macaulay wrote in 1800 in his *History of England*, "the smallpox was always present, filling the churchyard with corpses, tormenting with constant fears all whom it had not yet stricken, leaving on those whose lives it spared the hideous traces of its power, turning the babe into a changeling at which the mother shuddered, and making the eyes and cheeks of a betrothed maiden objects of horror to the lover." Particularly terrifying was the power of smallpox to destroy overnight the health and beauty of youth. In William Makepeace Thackeray's historical novel, *The History of Henry Esmond* (1852), the eighteenth-century narrator describes the ravages of the disease in the English countryside: "I remember in my time hundreds of the young and beautiful who have been carried to the grave, or have only risen from their pillows frightfully scarred and disfigured by this malady."

The pervasive impact of smallpox on human cultures around the globe was reflected in the widespread practice of worshiping gods, goddesses, and patron saints associated with the disease. Europeans throughout the Middle Ages revered St. Nicaise, a fifth-century bishop of Rheims who survived smallpox through holy practices a year before his martyrdom at the hands of the Huns. In Japan, a picture of Tametomo, a twelfth-century hero who killed a smallpox demon, was hung in the rooms of smallpox victims to aid their recovery. In West Africa, the Yoruba and neighboring tribes worshiped a smallpox deity called Sopona, and in Brazil, the god Omolou held sway over the disease. In China, Buddhists, Taoists, and Confucians alike revered the smallpox goddess T'ou-Shen Niang-Niang, and in northern India, Hindus erected numerous shrines and temples in honor of Shitala Mata.

One reason for the pervasive sense of fatalism and mystery about smallpox was that once a person contracted the disease, no effective treatment was available. Until the mid-nineteenth century, the sources of smallpox were variously ascribed to angry gods, noxious gases seeping from the earth, or imbalances of body humors. The causative agent of the disease was given the generic name *virus,* from the Latin word meaning "a poisonous force," but its true nature remained unknown. Rhazes, a Persian-born physician who lived in Baghdad around A.D. 910, was the first to distinguish the rash of smallpox from that of measles. He speculated incorrectly, however, that the human body harbored seedlike agents called "ferments" that were expelled through the skin to cause the pus-filled boils.

Rhazes recommended treating smallpox patients by bleeding them to the point of fainting and then exposing them to intense heat. Other popular folk remedies included palm oil, powdered horse excrement, and purgatives. One peculiar belief, originating in Japan in the tenth century and spreading to Europe in the twelfth century, was that the color red had therapeutic benefits because it served to excite the blood and bring the infection to the surface. Smallpox patients were accordingly wrapped in red blankets, dressed in red clothes, illuminated with red lamps, and had their rooms draped with red curtains. When Queen Elizabeth I of England contracted smallpox in October 1562, at the age of twenty-nine, her doctors cloaked her in a red blanket and placed her close to a roaring fire. The queen managed to survive the disease and died at the ripe old age of seventy, having ruled England for forty-five years. But her bout with smallpox left her bald and with disfiguring facial scars, and for the rest of her life she wore a red wig and heavy makeup.

Superstitious practices to ward off smallpox were also widespread. In eighteenth-century Europe, people tried to protect themselves from infection by holding vinegar-soaked rags over their noses, wearing amulets of animal teeth or bags of camphor around their necks, and carrying lengths of tarred rope. A more effective means of containing smallpox epidemics was quarantine. Long before the germ theory of disease was developed, people deduced that smallpox was contagious because the rash was so distinctive and the infection usually occurred after close contact with a sick individual. For this reason, smallpox patients in Europe and America were often placed in isolation on ships moored in harbors or in special quarantine hospitals. In the 1700s, for example, "pest houses" were built on two islands in Boston Harbor. Foreign ships approaching the city were required to stop and discharge passengers with smallpox or plague; these individuals were housed in dreary isolation wards until they either died or recovered from the illness.

Other than quarantine, the only effective means of combating smallpox was through deliberate infection. It had been observed in India and China that pockmarked survivors never again contracted the disease. This observation led to the idea of preventing natural smallpox by making small incisions in the skin of healthy people and inoculating them with scabs or pus from smallpox patients who had a mild form of the disease. Known as "buying the smallpox" or "variolation," the practice of smallpox inoculation began in India sometime before 1000 B.C., spread to Tibet, and was introduced into China by monks at a Buddhist monastery in Sichuan province around A.D. 1000.

For some unknown reason, introducing smallpox through the skin rather than the respiratory tract, the natural route of infection, had the effect of reducing the fatality rate from 30 percent to about 1 percent. In most cases, only a few dozen pustules appeared around the inoculation site, yet the resulting lifelong immunity was equivalent to that produced by the full-blown illness. Variolation entailed substantial risks—about one in a hundred people developed a fatal case of smallpox—but when faced with the near-certainty of contracting the natural disease, many willingly took the gamble.

Different forms of variolation were practiced in various parts of the world. A Chinese method, known as "insufflation," involved grinding dried smallpox scabs into a fine powder, which was then sucked into the nose through an ivory straw in the manner of taking snuff. Yu T'ien-Chih described this technique in his book *Miscellaneous Ideas in Medicine,* published in 1643. In Russia, recipients went to a bathhouse and had their skin slapped with branches that had previously been used on a smallpox victim. During the mid-seventeenth century, merchant caravans brought knowledge of variolation to Arabia, Persia, and North Africa, and it came to be practiced at the folk level throughout the Ottoman Empire, mainly by old women. The Turks used the method of skin inoculation, which they called "engrafting."

In the early eighteenth century, Lady Mary Wortley Montagu, a London socialite and wife of a member of Parliament, brought the practice of variolation to England. A famous beauty in her youth, she had contracted smallpox in December 1715 at the age of twenty-six, after her younger brother had died of the disease. She survived the ordeal, but at the cost of severe facial scarring and the loss of her eyelashes. In 1717, Lady Montagu accompanied her husband to Constantinople (now Istanbul), where he had been appointed British ambassador. Two weeks after her arrival, she became familiar with the practice of variolation at the Ottoman court and sent a letter to a friend describing the procedure. "Every year," she wrote, "thousands undergo this operation, and the French ambassador says pleasantly, that they take the small-pox here by way of diversion, as they take the waters in other countries."

Determined to protect her five-year-old son Edward from smallpox, Lady Montagu ordered the embassy surgeon, Dr. Charles Maitland, to variolate the boy in March 1718. The operation was performed while her presumably skeptical husband was away and despite the strong opposition of the embassy chaplain, who considered the procedure un-Christian. Fortunately, it was successful. After returning to London, she had Dr. Maitland variolate her three-year-old daughter in the presence of physicians from the royal court.

In 1721, despite resistance from the medical establishment, Lady Montagu persuaded the scientifically minded Princess of Wales and her husband to sponsor a public experiment to demonstrate the safety of variolation. This proposal fell on receptive ears because of the disastrous impact of smallpox on the British succession, including the death in 1700 of

Queen Anne's son and sole surviving heir. The Princess of Wales offered a full pardon to six condemned prisoners facing the gallows at Newgate Prison in London on the condition that they serve as subjects in what was called the Royal Experiment. The prisoners agreed, and in August 1721 they were variolated in front of an audience of physicians, surgeons, apothecaries, and members of the press.

All six subjects recovered from the procedure with no ill effects and were duly pardoned. To demonstrate the protective effects of variolation, one of the subjects, a nineteen-year-old woman, was ordered to sleep in the same bed with a ten-year-old smallpox victim for six weeks but did not contract the disease. Persuaded by this demonstration, the Princess of Wales had the king's surgeon inoculate her two young daughters, princesses Amelia and Caroline, on April 17, 1722. Thanks in part to the royal example, variolation became popular in England by the 1740s. The procedure was not widely accepted on the European continent, however, until after King Louis XV of France died suddenly of smallpox in 1774.

Variolation was often conducted in special "pox houses" or "inoculation hospitals" that opened in a number of countries. Most variolators required the recipient to undergo a lengthy preparation period of fasting, bleeding, and purging, which, it was believed, would ensure a milder case of the disease. Only the relatively affluent could afford the time and medical fees required for the procedure. In the 1760s, Daniel Sutton and Thomas Dimsdale developed a simplified form of variolation that required little or no preparation of the patient and reduced the size of the incision and the amount of inoculum to diminish the risk of complications. This modified technique usually produced an illness that was considerably milder than natural smallpox, resulting in only a few pustules at the inoculation site and a much lower risk of death.

Despite such improvements, however, variolation was widely opposed by the medical establishment as unsafe and by the church as an interference with God's will. It did seem illogical to attempt to preserve people's health by making them sick, and the procedure was undeniably risky: King George III's son Octavius died as a result of variolation in 1783. On other occasions, recipients became infected with other fatal diseases, such as tuberculosis or syphilis, carried by the pus of the donor. Moreover, although people who had been variolated often acquired an extremely mild case of smallpox and felt well enough to resume their normal activities, they could still transmit the disease through the air to infect others. Thus, in the absence of strict quarantine, variolation often spawned additional epidemics. These hazards led many communities to ban the procedure outright.

Variolation was less popular in major cities such as London, where smallpox was largely a childhood disease and the poor were burdened by an excess of offspring, than in small towns and rural villages, where more sporadic outbreaks struck down adolescents and young adults. The procedure also became established in the British colonies of North America, whose isolated nature made them vulnerable to severe epidemics. In 1721, for example, smallpox broke out in Boston, sickening 5,980 of the city's 11,000 residents and killing 844. The Reverend Cotton Mather, who had learned of variolation from his African slave Onesimus, persuaded local physician Zabdiel Boylston to practice the technique. Of the 242 people who were inoculated, only six died—a much lower mortality rate than that inflicted by the natural disease. But the Boston medical establishment, the clergy, and much of the general public were bitterly opposed to variolation. Boylston's life was threatened and he was forced to go into hiding.

In 1722, Reverend Mather attempted to popularize variolation in the American colonies by publishing a small book titled *An Account of the Method and Success of Inoculating the Small-Pox,* which gave a step-by-step description of the procedure. During a subsequent epidemic in Boston in 1753–54, Benjamin Franklin, who had lost a son to smallpox, conducted a scientific study to demonstrate the effectiveness of variolation and became an enthusiastic supporter of the technique. In Russia, variolation was popularized by Catherine the Great, who was inoculated along with her son in 1768. She then issued orders to variolate her subjects, and numerous "pox houses" were established for this purpose.

Although variolation could protect against infection, it could also be misused to trigger deliberate outbreaks of smallpox. The fact that isolated populations, such as American Indians, were highly susceptible to the disease made it a potential weapon in the hands of less susceptible groups, such as Europeans. Indeed, in a dark chapter of military history, the British employed smallpox as an instrument of warfare on several occasions during the eighteenth century.

The best-documented incident occurred in the aftermath of the French and Indian War of 1754–63, when Great Britain defeated France and its allied Indian tribes and seized control of Canada. After the war ended, Pontiac, an Ottawa chief who had sided with the French, was angered by the British confiscation of Indian land. Seeking to drive the British out of Canada and the Mississippi watershed and return the territory to French control, Pontiac united six tribes along the western frontier into a military alliance.

Because most of the British army had returned home, the remaining units were badly overextended. As a result, the rebellious Indians, led by Pontiac, overran eight British forts in western Pennsylvania, killing or capturing the defending soldiers and settlers. On May 29, 1763, the Delaware, Shawnee, and Mingo tribes began a siege of the major outpost of Fort Pitt (site of present-day Pittsburgh), which soon was seriously threatened. Indian scalping parties attacked British settlements around the fort, destroying harvests, butchering men, women, and children, and forcing the survivors to flee in terror.

Colonel Henry Bouquet, the ranking officer for the Pennsylvania western frontier, headquartered in Philadelphia, wrote a letter on June 23, 1763, describing the increasingly dire military situation at Fort Pitt. The letter was addressed to Sir Jeffrey Amherst, the British commander-in-chief in North America, based in New York. In addition to describing the Indian attacks, Bouquet reported that smallpox had broken out in the defending garrison.

On July 7, 1763, Amherst responded, adding a postscript to his letter in which he suggested that the defenders of Fort Pitt should use smallpox as a weapon against the tribes involved in Pontiac's Rebellion. "Could it not be Contrived to Send the *Small Pox* among those Disaffected Tribes of Indians?" he wrote. "We must, on this occasion, Use Every Stratagem in our power to Reduce them." In response to Amherst's recommendation, Bouquet replied elliptically on July 13, "I will try to inoculate _____ by means of Some Blankets that may fall in their Hands, taking care however not to get the disease myself." Amherst responded approvingly on July 16, noting, "You will Do well to try to Inoculate the Indians by means of Blanketts, as well as to try Every other Method that can serve to Extirpate this Execrable Race."

Although the available documents do not reveal whether Bouquet carried out Amherst's suggestion, officers at Fort Pitt had already taken the initiative and executed a similar plan a few months earlier. William Trent, the commander of the local militia, wrote in his journal on May 24, 1763, that when a small delegation of Delaware Indians had visited the fort to advise the British to surrender, he had given them "two Blankets and a Handkerchief out of the Small Pox Hospital. I hope it will have the desired effect." Captain Simeon Ecuyer, the commanding officer at Fort Pitt, was aware of this operation, since he subsequently approved Trent's invoice to replace the blankets and the handkerchief. A severe epidemic of smallpox subsequently broke out among the Indians besieging Fort Pitt in the summer of 1763, but whether it resulted from Trent's operation or from natural causes remains unknown.

Allegations of the deliberate use of smallpox as a weapon also surfaced during the American Revolutionary War. During the early years of the conflict, British troops who had not already experienced the disease were routinely variolated, but the colonists were not. On several occasions, the Americans accused the British of deliberately spreading smallpox. In April 1775, for example, the Continental army, under the command of General George Washington, besieged the British forces encamped in Boston. When smallpox broke out in the city in December 1775, British General William Howe ordered the variolation of all susceptible troops as a protective measure. Reportedly, however, the British also inoculated civilian refugees leaving the city in a deliberate attempt to spread the disease among the American forces. Although General Washington initially did not believe the rumors of this diabolical scheme, he changed his mind a few days later when smallpox broke out among Bostonians fleeing the British. The Continental army managed to contain the epidemic by quarantine and disinfection, but Washington's fear of exposing his army to smallpox led him to delay the liberation of Boston. When the British troops finally evacuated in March 1776, he ordered one thousand men who had already survived smallpox to secure the city. Despite this precaution, a large number of troops and civilians fell victim to the disease.

During the winter of 1775–76, another siege was under way at Quebec City. Continental forces led by Benedict Arnold were striving to free Quebec from British control and add it to the territory of the thirteen colonies. Having captured Montreal, the Americans were poised to take Quebec City in early December 1775, when the British fort commander reportedly had civilians variolated and sent out to mingle with the Continental troops. A few weeks later, a major smallpox epidemic broke out in the American ranks, affecting roughly half the ten thousand soldiers. After burying their dead in mass graves, the Continental army retreated in disorder. Thomas Jefferson later alleged that smallpox had been "sent into our army designedly by the [British] commanding officer in Quebec." Were it not for that epidemic, Quebec and perhaps all of Canada might be part of the United States today.

Historian Elizabeth Fenn has also discovered evidence that the British attempted to transmit smallpox to Continental forces in some of the southern states. Because many of the American revolutionaries were slaveholders, black slaves who escaped during the war often joined the British army in return for the promise of freedom. In May 1776, the British royal governor of Virginia sent smallpox-infected slaves back to their home plantations in an apparent attempt to spread the disease.

In 1777, as a result of repeated outbreaks of smallpox among the colonial troops, General Washington ordered the entire Continental army variolated before he launched any new military operations. All soldiers who had not experienced smallpox were inoculated at special hospitals established for this purpose. Washington, who at age nineteen had survived a case of smallpox that had pockmarked his face, considered the disease to be his most dangerous enemy.

DR. JENNER'S VACCINE

The practice of variolation continued until Edward Jenner, an English country doctor in the town of Berkeley, Gloucestershire, discovered a safer alternative. Jenner himself had been variolated in 1757 at the age of eight, after a preparatory period involving six weeks of fasting and bleedings. The procedure had made him severely ill and he had required months of recuperation—a deeply traumatic experience that had persuaded him of the need for a better approach to smallpox prevention.

In addition to his successful medical practice, Jenner devoted his spare time to the study of natural history, and published monographs on hedgehogs and cuckoos that earned him membership in the British Royal Society. His powers of observation led him to notice that milkmaids in the English countryside had unusually fair complexions and seemed largely immune from the disfiguring effects of smallpox. In 1770, a milkmaid told him that contracting cowpox had protected her from smallpox. Although the medical establishment frowned on such folk wisdom, Jenner thought it might have some merit. Milkmaids often became infected with cowpox, a sporadic disease of dairy cattle that produced sores on cows' udders. In humans, cowpox caused a few bluish pustules on the milker's hands but no other serious symptoms.

Jenner speculated that exposure to the milder condition had somehow shielded the girl against the far more deadly disease. Over the next several years, he variolated thirteen adults who had never had smallpox but had previously contracted cowpox or horsepox, a related infection. None of the recipients developed a skin rash at the inoculation site, indicating that they were already immune to smallpox. During an outbreak of cowpox in the English countryside in May 1796, Jenner went one step further: He sought to determine whether deliberate inoculation with cowpox could induce immunity to smallpox in a child who had never experienced either disease. On May 14, in a bold experiment that would be considered unethical by today's standards, Jenner made two half-inch incisions in the arm of James Phipps, the eight-year-old son of a laborer who often worked for him, and inserted pus and lymph extracted from a cowpox pustule on the hand of a young milkmaid named Sarah Nelmes. After three days, Phipps developed swelling, redness, and blister formation at the site of inoculation. The lesion scabbed over and healed in two weeks, leaving a small scar. One month later, on July 1, Jenner inoculated the boy with pus from a case of smallpox.

In a letter to a friend, the intrepid doctor described the results of his experiment. "But now listen to the most delightful part of my story. The boy has since been inoculated for the Smallpox which, as I ventured to predict, produced no effect. I shall now pursue my Experiments with redoubled ardor." Despite this laconic description, Jenner had made a momentous discovery: Inoculating a nonimmune person's skin with material from a cowpox lesion had induced immunity against smallpox. Cowpox inoculation therefore offered the protective effects of variolation but without the considerable risks. Although James Phipps reportedly suffered from a tubercular hip or spine, he lived into old age, and in 1818 Jenner built him a house in Berkeley as a token of appreciation.

Because Jenner's research preceded the germ theory of disease and the discovery of virus particles, his approach was entirely empirical. He had no idea why a mild infection with cowpox protected against a much more virulent one with smallpox. Today we know that the procedure worked because of similarities between cowpox and variola at the molecular level. The two viruses are close cousins: different species within the same genus that probably originated from a common progenitor. After Jenner inoculated James Phipps with pus containing cowpox particles, the virus began to replicate in the boy's skin, stimulating his immune system to produce antibodies and white blood cells targeted against some of the proteins in the viral coat. The cowpox virus was mild enough for Phipps's immune system to keep it in check, so the infection remained localized in the skin near the inoculation site and did not spread throughout his body. Because of the similarity between the coat proteins of cowpox virus and those of variola, however, the antibodies and white blood cells that his immune system produced to counter the cowpox infection also protected him against the subsequent inoculation with smallpox.

In 1797, Jenner submitted a paper to the British Royal Society describing his findings, including the results of the "challenge" experiment with James Phipps. Sir Joseph Banks, the president of the Royal Society, declined to publish the manuscript on the grounds that Jenner's ideas were too revolutionary and his experimental proof too limited. Banks advised the author that he ought not to risk his reputation by presenting to the learned body findings that appeared so much at variance with established knowledge.

Despite this setback, Jenner did not abandon his theory but resolved to collect more evidence. The absence of cowpox from local herds precluded him from further experimentation until March 1798, when an outbreak at a nearby farm enabled him to resume his studies. At that time, he gave cowpox inoculations to several more children who had never had smallpox, including his eleven-month-old son, Robert. After the vaccination pustules had healed, Jenner variolated four of the subjects and found that all were immune to challenge with smallpox.

Having revised and expanded the experimental section of his manuscript, Jenner decided, on the advice of friends, to publish the paper privately at his own expense. The sixty-four-page pamphlet finally appeared in 1798 under the title, *An Inquiry into the Causes and Effects of* Variolae Vaccinae, *a Disease Discovered in Some of the Western Counties of England, Particularly Gloucestershire, and Known by the Name of the Cow Pox.* To describe the cowpox inoculum, Jenner coined the word *vaccine* from the Latin word *vacca,* for "cow," and he named the process "vaccination."

The discovery of vaccination marked a turning point in medical history and a fundamental change in humanity's relationship to disease. For the first time, it was possible to take a harmless measure to prevent a deadly infection before it occurred. In 1881, the great French microbiologist Louis Pasteur honored Jenner by broadening the use of the term "vaccine" to refer to *any* inoculated material that induced immunity against an infectious disease. Nevertheless, Jenner's use of cowpox virus to protect against smallpox was different from the vaccines developed later to combat other viral diseases, such as polio, measles, and yellow fever. The latter vaccines were either killed versions of the original disease-causing virus or living strains that had been genetically weakened, or "attenuated," so that they no longer caused clinical illness.

Vaccination offered several important advantages over variolation. Recipients of cowpox vaccine did not contract smallpox and hence could not transmit the disease to others. Moreover, the side effects of vaccination, such as fever, were usually quite mild. The significance of Jenner's discovery was recognized almost immediately and the news spread quickly. Within a year, at least a thousand people had been vaccinated and the *Inquiry* had been translated into German, French, Dutch, and Latin.

Nevertheless, Jenner's pamphlet sparked a firestorm of criticism from the clergy and the medical establishment, who argued that inoculating a healthy person with material from a diseased animal was repulsive, unsafe, and ungodly. A satirical cartoon published in 1802 in a London newspaper showed the recipients of cowpox vaccine sprouting bovine horns and tails. Other critics sought to discredit Jenner by claiming that he had not been the first to demonstrate the protective power of cowpox. They pointed out that in 1774, some twenty-two years before Jenner performed his challenge experiment on James Phipps, a cattle breeder named Benjamin Jesty had inoculated his wife and sons with cowpox in an effort to protect them against smallpox. Jesty, however, had made no attempt to publicize what he had done, whereas Jenner had engaged in systematic experimentation and had published his findings widely.

Jealous of Jenner's discovery, a few prominent English physicians adopted his technique without proper attribution. William Woodville, the head of London's Small Pox and Inoculation Hospital, obtained cowpox material during a local outbreak in cattle and vaccinated several hundred persons. Unfortunately, some of the vaccine that Woodville supplied to other physicians was contaminated with smallpox, inadvertently spreading the disease. In response, Jenner published a second edition of his pamphlet in which he warned against inadvertent contamination and stressed the importance of using the vaccine before its protective character was lost.

The new vaccination technique also found its way to the United States. In 1799, a year after Jenner's *Inquiry* was published, Dr. Benjamin Waterhouse, one of the first three professors at Harvard Medical School, obtained a copy of the pamphlet and was deeply impressed by it. "On perusing this work," he wrote in 1800, "I was struck with the unspeakable advantages that might accrue to this country, and indeed to the human race at large, from the discovery of a mild distemper that would ever after secure the constitution from that terrible scourge, the smallpox." Waterhouse obtained a sample of cowpox from England and administered it to his five-year-old son Daniel and several other family members and servants, the first vaccinations performed in the United States. He then tested the effectiveness of the vaccine by sending a twelve-year-old servant boy he had vaccinated to the Small-pox Hospital in the neighboring town of Brookline. The boy returned home unharmed after several days and remained free of the infection thereafter.

Waterhouse wrote a pro-vaccination pamphlet titled *On the Prospect of Exterminating the Smallpox* and enlisted the help of Vice President Thomas Jefferson, who became an ardent supporter of the technique. "Every friend of humanity must look with pleasure on this discovery, by which one more evil is withdrawn from the condition of man," Jefferson wrote to Water-house in 1800. The following year, Jefferson vaccinated two hundred of his Virginia neighbors and arranged to provide smallpox vaccine to several Indian tribes.

Waterhouse's advocacy of vaccination aroused the ire of the conserva-tive Boston medical establishment. In 1802, a group of local physicians petitioned the city's board of health to conduct a comparative trial of vac-cination and variolation. To their chagrin, the experiment clearly demon-strated the superiority of vaccination, which the board then endorsed. But Waterhouse aroused further criticism by refusing to share his vaccine with certain local physicians and by demanding a share of the profits from others.

By the early 1800s, more than 100,000 persons in Great Britain had been vaccinated. Although the world was clamoring for Jenner's vaccine, it was difficult to obtain an adequate supply because outbreaks of cowpox in Eu-ropean cattle were sporadic and unpredictable and the disease did not exist on other continents. The vaccine consisted of the "matter" of cowpox—pus and lymph containing the live virus—which had to be extracted from pustules on the udders of infected cows at just the right stage of the dis-ease. Since the cowpox material had to be "active" for vaccination to be effective, distribution and handling posed major hurdles. Although the vac-cine could be preserved for a few months by drying it on threads, quills, glass, or slivers of ivory, it was rapidly inactivated by high temperatures or by exposure to sunlight, factors that were poorly understood at the time.

As a result, cowpox material shipped over long distances often lost its effectiveness before arriving at its destination.

One solution to this problem was to keep the vaccine "alive" by transferring it from one human recipient to the next, a practice known as the "arm-to-arm" technique. First, an individual was vaccinated, and as soon as a cowpox pustule had appeared on his or her arm, matter from the lesion was then used to vaccinate other recipients. In 1801 in St. Petersburg, Russia, for example, a recently vaccinated girl was sent to a local orphanage to serve as the source of smallpox vaccine for all children more than a week old. From then on, the orphanage continuously transferred the vaccine from one child to another for more than ninety-two years (1801–93).

The arm-to-arm method was also used to distribute Jenner's vaccine throughout the Spanish Empire. Spanish King Charles IV's daughter had been stricken with smallpox in 1798, and after she recovered, he arranged to have the rest of his family vaccinated. In 1803, the king, convinced of the benefits of the vaccine, ordered his personal physician, Francis Xavier de Balmis, to deliver it to the Spanish dominions in North and South America and, if possible, in Asia as well. To maintain the vaccine in a viable state during the long sea voyage, the physician recruited from the orphanages of Spain twenty-two young boys, aged three to nine years, who had never had cowpox or smallpox before.

During the trip across the Atlantic, de Balmis sequentially vaccinated the orphans in a living chain. Two children were vaccinated immediately before departure, and when cowpox pustules had appeared on their arms, material from these lesions was used to vaccinate two more children. This procedure was repeated at roughly ten-day intervals until the ships arrived in Venezuela so that at least one child always had a cowpox pustule at the right stage of maturation to provide the active vaccine. In return for the orphans' service, the Spanish royal court arranged foster families for them and paid their living and schooling expenses in the New World. Thanks to the Royal Smallpox Expedition, more than 100,000 people in Latin America were vaccinated. The expedition then recruited twenty-six more orphans and continued transferring the vaccine from arm to arm throughout the remainder of the three-year voyage, which included stops in the Spanish Philippines, Macao, and Canton.

The arm-to-arm method was not always practicable, however, and other means of transporting the vaccine over long distances were limited. Samples of cowpox dried on pieces of cotton thread and stored at room temperature had a shelf life of only three months. Since transportation was slow and refrigeration nonexistent, the vaccine often lost its potency by the time it was administered. In 1803, for example, President Thomas Jefferson sent Meriwether Lewis and William Clark to explore the Louisiana Purchase, which he had just acquired from Napoleon, and gave them a supply of cowpox to vaccinate the Indians they encountered. Although Lewis and Clark found tribes in remote areas that had been decimated by smallpox, the cowpox vaccine they carried with them soon lost its potency, and the replacement supply they requested was never delivered.

The initial enthusiasm over vaccination also waned when some people who had been successfully vaccinated in childhood contracted smallpox as adults. Until his death in 1823, Jenner continued to claim that his vaccine provided lifelong protection, as was the case with variolation, but it gradually became apparent that this was not the case. Jenner explained the increasing number of smallpox cases in adults who had been vaccinated in childhood as the result of "imperfect vaccination" with viruses that appeared to be cowpox but were not. Although it was true that diseases other than cowpox could cause lesions on the teats of cows, the fact was that the immunity induced by vaccination diminished over time, depending on the frequency of exposure to natural smallpox. On average, effective immunity lasted about seven to ten years, although it varied markedly among individuals. Those who had been vaccinated at least once had a lower risk of infection compared with the never-vaccinated, but vaccination had to be repeated on a regular basis to ensure a high level of protection. People who had been vaccinated once but whose immunity had declined could still contract smallpox, but they often acquired a milder, less recognizable form of the disease.

During the first half of the nineteenth century, several European countries adopted the practice of vaccination. In 1805, Napoleon Bonaparte ordered that all French troops who had not already had smallpox should receive Jenner's vaccine. By 1821, the routine vaccination of infants was compulsory in Bavaria, Denmark, Norway, Bohemia, Russia, and Sweden, resulting in a rapid decline in the incidence of the disease. After a major smallpox epidemic took place in Britain in 1837–40, Parliament passed the Vaccination Act of 1840, which outlawed variolation and mandated the vaccination of infants free of charge.

Despite the growth of vaccination, smallpox remained pervasive in Victorian England, particularly among the impoverished classes. Charles Dickens portrayed the disease in his novel *Bleak House,* published in 1853. Esther, the heroine of the story, has a young maid named Charley who takes in a homeless boy suffering from high fever and back pain, the early symptoms of smallpox. Charley subsequently contracts the infection from the boy, and Esther nurses her back to health. During the maid's illness, Esther writes in her diary, "I was very sorrowful to think that Charley's pretty looks would change and be disfigured, even if she recovered—she was such a child, with her dimpled face—but that thought was, for the greater part, lost in her greater peril."

In the United States, the practice of vaccination experienced numerous vicissitudes. President James Madison, who served from 1809 to 1817, named Dr. James Smith of Maryland as the federal agent responsible for distributing smallpox vaccine. Public opposition to vaccination remained strong, however, and in the early 1820s, a shift in the political climate led the U.S. Congress to repeal the smallpox vaccination law and remove Dr. Smith from office. Soon, the incidence of the disease was again on the rise in the United States.

During the American Civil War, smallpox inflicted casualties on both sides, but the Confederates relied more heavily than the Union forces on arm-to-arm vaccination and suffered the drawbacks of that technique. Before the Battle of Chancellorsville on May 1–3, 1863, the vaccine used by the Confederate troops became contaminated with syphilis, taking some 5,000 soldiers out of action. Later that year, President Abraham Lincoln contracted smallpox and was feverish with the early phase of the disease when he gave his famous Gettysburg Address on November 19. Fortunately, he had a mild case and recovered with little facial scarring.

Some evidence suggests that on a few occasions during the American Civil War, smallpox may have been used deliberately as a biological weapon. Dr. Luke Blackburn, who later became the governor of Kentucky, allegedly contaminated clothing with smallpox and sold it to unsuspecting Union troops. The obituary of at least one Union officer stated that he had died of an outbreak attributed to Blackburn.

The continuing shadow of smallpox is evident in "The Schooner Maryann," a sea shanty from around 1865, which describes events on board a merchant ship bound from New York City to Newfoundland. Three days after the ship sets sail, the thirty-year-old captain, a "strapping youth," falls gravely ill. He dies a few days later and is buried in Nova Scotia. The last two stanzas read as follows:

> The doctor he was called on board
> His death for to make known,
> Smallpox on board was raging,
> Was told to every man,
> It was on the following evening
> Two more were sent on shore;

May the Lord have mercy on their souls,
We shall never see them more.

Out of five bold youths that left New York
Only two did now return
Home to their wives and families
Their losses for to mourn.
Home to their wives and families,
And never more to roam,
And learn to live as landsmen do
Forever safe at home.

A major obstacle to the effective prevention of smallpox was that many of the technical problems with the vaccine remained unsolved. In addition to the difficulty of keeping cowpox alive, it could be hard to distinguish from other viruses that caused lesions on cow udders, resulting in ineffective vaccines. The arm-to-arm method could also contaminate the vaccine with dangerous pathogens, resulting in the inadvertent spread of hepatitis or syphilis. In 1861, for example, forty-one Italian children who had been vaccinated by arm-to-arm transfer acquired syphilis from a child with an undiagnosed case of the disease.

Clearly, a better and safer method of obtaining the vaccine was needed. As early as 1805 in Naples, Italy, it had been discovered that cowpox could be grown in large quantities by making a series of deep scratches in the flank of a calf, inoculating the virus into the animal's skin, and harvesting the pustules five to ten days later. This method first became known outside Italy when it was presented at the Medical Congress of Lyons, France, in 1864. Thereafter, production of vaccine in the scarified skin of calves (and sometimes sheep or water buffalo) was gradually adopted in other countries. The crude smallpox vaccine grown in calf skin was far from ideal, however.

Essentially a mixture of pus and lymph containing cowpox virus, it was inevitably contaminated with animal hair, cells, bacteria, and other viruses. There was no regulation of vaccine production or quality, and without refrigeration, the vaccine deteriorated rapidly at warm temperatures. Because of these limitations, smallpox remained a fact of life even in countries where vaccination was practiced widely.

The European continent was slow to make vaccination mandatory, and at least a half-million soldiers and civilians died in the smallpox epidemic associated with the Franco-Prussian War of 1870–71. Most of the victims were French because of the widespread belief in that country that revaccination was not required for continued protection. After the war, the French Parliament passed a law imposing fines, and even imprisonment, on parents who failed to have their children vaccinated against smallpox. At the same time, antivaccination societies proliferated in Europe, with more than one hundred groups organizing protests by 1889. Opponents of vaccination questioned its safety and efficacy and believed that compulsory vaccination was a violation of civil liberties. Among those opposed to vaccination was the British playwright George Bernard Shaw, who called it "a particularly filthy piece of witchcraft."

During the late nineteenth century, pioneering scientists obtained new insights into the nature of smallpox and other viral infections. In 1878–80, Louis Pasteur developed the germ theory, which, for the first time, implicated microscopic organisms as the causative agents of infectious diseases. This breakthrough was followed by Robert Koch's identification, in 1882–83, of the bacteria that give rise to tuberculosis and cholera. Then, in 1890, the Russian microbiologist Dimitry Ivanovsky obtained experimental proof for the existence of viruses: infinitesimally small infectious agents that are invisible under the light microscope, even under great magnification, and that can penetrate porcelain filters designed to trap ordinary bacteria. (Because viruses are about one-hundredth the size of bacteria, they were not seen in any detail until 1937, after the electron microscope had been invented.)

Despite the new scientific understanding of smallpox, effective treatments remained elusive. Popular nineteenth-century remedies for smallpox included vegetarianism, meditation, and patent medicines. During an epidemic of smallpox in Muncie, Indiana, in 1893, the local newspapers published two "infallible" cures. One involved dissolving an ounce of cream of tartar in a pint of boiling water and drinking it at intervals during the day; the second called for mixing a grain of sulphate of zinc and a grain of foxglove with four ounces of water, and taking a teaspoon of this solution every hour. Needless to say, neither treatment was effective.

During the summer of 1897, variola minor or alastrim, the milder form of smallpox, appeared in Pensacola, Florida. Within four years, it had spread across the United States and had been imported into some countries of South America. Although sporadic outbreaks of virulent smallpox occurred in U.S. cities until about 1927, most cases after that year were alastrim, which had a case fatality rate of less than 1 percent. As a result, interest in vaccination waned. Although the U.S. Supreme Court had ruled in 1922 that states had the right to require smallpox vaccination for school admission, antivaccination sentiment and resistance to compulsory measures remained high in many parts of the country. By the 1930s, only nine states and the District of Columbia had mandatory vaccination laws, and four states explicitly prohibited compulsory vaccination.

A key breakthrough in the control of smallpox in industrialized countries was the invention of the icebox, which made it possible to refrigerate smallpox vaccine and thereby preserve its potency for long periods. Thanks to this innovation and to mandatory child vaccination, the reported incidence of smallpox in the United States declined from 102,791 cases in 1921 to between 5,000 and 15,000 cases annually from 1932 to 1939. World War II, however, caused a major setback to the worldwide control of the disease. Over the period 1941 to 1946, the number of countries in which smallpox was endemic, or continually present, rose from sixty-nine to eighty-seven.

Smallpox continued to circulate in the United States as late as 1946, and perhaps 1947. Although 357 cases were reported in 1946, the reporting was not more than 5 percent to 10 percent complete. In addition, several cases of virulent smallpox were imported into the country from abroad. In 1945, an American soldier returning from Japan triggered an outbreak in Seattle totaling sixty-five cases, with twenty deaths. In March 1947, a businessman traveled by bus from Mexico to New York City, where he was hospitalized with a severe case of smallpox and died five days later. By that time he had infected four people, who in turn infected seven more, resulting in a total of twelve cases and two deaths. This outbreak triggered near-hysteria in New York City and led to the emergency vaccination of more than six million people in less than a month, successfully containing the epidemic but completely depleting the city's vaccine supply. The last smallpox outbreak in the United States occurred in February 1949, when an individual from Mexico imported the disease into the Rio Grande Valley of Texas, causing seven secondary cases and one death.

Despite the conquest of smallpox in the United States and Europe, the disease continued to rage unabated in much of the developing world, particularly in tropical countries where the refrigeration of smallpox vaccine was impractical. Because the liquid vaccine was not heat-stable, it deteriorated rapidly unless it was maintained at a temperature of 41 to 50 degrees Fahrenheit from the time the vaccine was produced until it was inoculated into a human subject, a technical requirement known as the "cold chain." Some attempts were made to keep the vaccine chilled by transporting it in portable kerosene refrigerators on muleback, but this approach was exceedingly cumbersome and vaccine spoilage remained a major problem. As a result, vaccination induced effective immunity in fewer than 30 percent of recipients in tropical countries.

In the 1930s, Dutch and French scientists perfected air-dried vaccine preparations that remained potent at room temperature for as long as a month and could be rehydrated shortly before use. Although often heavily contaminated with bacteria and of variable potency, such air-dried vaccines

were used successfully in Dutch Indonesia and in the French colonies of
Africa and Asia. A further technical breakthrough came in the early 1950s,
when Leslie H. Collier at the Lister Institute in London developed a procedure for manufacturing stable, freeze-dried smallpox vaccine on a commercial scale. Collier published his method and made it freely available to
all laboratories that wished to use it. Vaccine technology had now progressed to the point where smallpox eradication had become a feasible
proposition.

After smallpox vaccine had been in widespread use for many years, a remarkable discovery was made: The vaccine strain being employed around the world
was not the cowpox virus that Jenner had used, but an entirely different
orthopoxvirus that did not exist in nature and became known as "vaccinia."
In 1939, Allan Downie of the University of Liverpool in England determined
that vaccinia was genetically distinct from both variola and cowpox. For decades, vaccinators had experimented with different vaccine strains, selecting
those that gave good protection but caused less fever and smaller skin lesions.
At some unknown point in time, they had begun using vaccinia instead of
cowpox. Where vaccinia virus had come from, and when it had become the
primary virus used to vaccinate against smallpox, remained a mystery.

Several competing hypotheses were proposed to explain the origin of
vaccinia. In 1967, British poxvirologists Keith Dumbell and Henry S. Bedson
suggested that vaccinia was a hybrid of cowpox and variola viruses that had
arisen from their inadvertent mixing during early vaccination programs.
In 1980, Peter Razzell argued that the vaccine used by Jenner and his con-
temporaries had not been cowpox at all, but rather a weakened strain of
variola that had gradually evolved into vaccinia during thousands of arm-
to-arm transfers. Derrick Baxby, in his 1981 book *Jenner's Smallpox Vaccine,* suggested that vaccinia may have evolved from horsepox (also known
as "grease"), which was used as a smallpox vaccine in some European countries during the nineteenth century and later became extinct in nature; the

clinical superiority of horsepox as a vaccine may have led to its adoption in place of cowpox. Because none of the hypotheses for the origin of vaccinia could be proven conclusively, however, the mystery remained unsolved.

Despite the lack of a drug treatment or cure for smallpox, Edward Jenner understood that widespread vaccination had the potential to eradicate the disease by depriving it of susceptible human hosts. In 1801, he predicted: "It now becomes too manifest to admit of controversy, that the annihilation of the Small Pox, the most dreadful scourge of the human species, must be the final result of this practice." Thomas Jefferson also subscribed to this view. In 1806, he wrote in a letter to Jenner, "Yours is the comfortable reflection that mankind can never forget that you have lived. Future nations will know by history only that the loathsome smallpox existed and by you has been extirpated."

In fact, 180 years were to elapse between Jenner's discovery of vaccination and the global eradication of smallpox. The missing elements were not scientific knowledge or technical capability, but rather the logistical infrastructure and international political will needed to undertake such a vast and costly campaign.

Four

LAUNCHING THE CRUSADE

Donald Ainslie Henderson, better known by his initials D. A., was born on September 7, 1928, in Lakewood, Ohio, a suburb of Cleveland. He grew up in a family of strict Scottish Presbyterians who stressed religion and education. Although his father was an engineer, D. A. decided at an early age to become a doctor, inspired by a physician uncle who was the senior member of the clan. At Oberlin College, he majored in liberal arts and took the required courses for medical school, while devoting much of his free time to extracurricular activities.

Tall and lanky (six feet, two inches), D. A. had a booming voice, an infectious grin, and an entrepreneurial spirit. He became editor of the college yearbook in his junior year, and he and his roommate founded a radio station that broadcast through the pipes of the campus heating system. They also devised creative strategies for bending the school rules. Although students were banned from having cars and everyone rode bicycles, there was no rule against motor scooters, which Henderson and his roommate used to zip around campus. Oberlin subsequently banned scooters but agreed to "grandfather in" the two young men on grounds of financial hardship. Exploiting another loophole, Henderson purchased a 1937 Oldsmobile as the "official car" of the college radio station. These experiences taught him the art of working around the regulations to get things done, a lesson he would later apply in his dealings with international bureaucracy.

Having discovered that he had a talent for managing people, D. A. decided to pursue a career combining medicine and administration. After graduating from Oberlin in 1950 with a liberal arts degree, he attended medical school at the University of Rochester, where he found that he identified more with the administrators than the attending physicians. He even made an appointment with the dean of the medical school to ask him about the career path for becoming dean. The man was taken aback; no student had ever asked him that question before. When Henderson completed his internship in 1955 at a hospital in Cooperstown, New York, he planned a residency in internal medicine followed by a specialization in cardiology. But the arrival of a letter from his local draft board led him in an unexpected direction.

Facing two years of peacetime military service, D. A. dreaded having to spend his days doing entry-level physicals. He learned that medical doctors had another option: working for the U.S. Public Health Service's Communicable Disease Center (CDC) in Atlanta. The catch was that Henderson was not particularly interested in infectious diseases. His encounters with the field during his medical training had mainly involved skin rashes in young children, from whom getting a history or taking a blood sample was problematic at best. Worse, the rashes had never looked like the pictures in the medical textbooks.

Despite Henderson's reservations about the field of infectious disease, the CDC position offered him the chance to spend his two years of military service doing something more interesting than routine physicals. There were other attractions as well: Public Health Service officers did not have to go to boot camp, wear uniforms, or salute, and the job involved a good deal of administration. With these advantages in mind, Henderson signed up. It would prove to be a fortuitous decision.

After moving to Atlanta, D. A. was inducted as an officer in the Epidemic Intelligence Service (EIS). This unit of the CDC had been created in 1951 by Dr. Alexander Langmuir, a Harvard-trained physician who had taught public health at Johns Hopkins University. At that time, the Korean War was raging and the U.S. government feared that enemy saboteurs might launch covert biological warfare attacks against military and civilian targets in the United States. Several types of disease agents were thought to be plausible weapons that could be released in the heart of a city by a lone operative with a briefcase. Although preventing such an attack seemed impossible, early detection and identification of the resulting epidemic would help to contain its spread and limit the number of casualties.

The first EIS class, which convened in July 1951, consisted of twenty-one physicians and one sanitary engineer who came to Atlanta for a four-week course in practical epidemiology, biostatistics, and public health administration. At the end of the training period, EIS officers were assigned either to CDC headquarters or to state public health departments and universities around the country. They remained on twenty-four-hour call to investigate suspicious epidemics, determine whether the cause was enemy action, and respond immediately with the appropriate medical treatment. After the Korean War ended in 1953, the primary mission of the EIS shifted away from biological warfare defense to the investigation and containment of natural outbreaks of infectious disease throughout the United States.

When Henderson arrived at the CDC in the summer of 1955, he attended the one-month EIS training course in practical outbreak investigation, or "shoe-leather epidemiology." Although his fellow officers were all draftees, they were a special breed: young, well educated, idealistic, and adventuresome. The EIS also differed from most government bureaucracies in its informal style and scientific orientation. Dr. Langmuir emphasized academic rigor and expected that each officer would produce at least one publishable paper during his two-year tour.

Henderson applied and was hired for the position of assistant chief EIS officer, basically a gofer for the senior person running the operation. When D. A. returned from his first field assignment investigating an outbreak of diphtheria, he found the chief EIS officer cleaning out his office. "I've been appointed health commissioner for the state of Alabama," the man explained.

"Well, what am I?" Henderson asked.

"You're the new chief EIS officer."

"But I don't know what I'm doing."

"You'll learn."

In short order, Henderson found himself running the surveillance section and reporting directly to Dr. Langmuir. With almost no formal training in epidemiology, he was taking emergency calls and directing EIS officers in the field. In this sink-or-swim situation, he learned quickly and under considerable duress. Langmuir was brilliant, charismatic, and inspirational to work for, but he was also a difficult and demanding man who did not tolerate incompetence and could be extremely harsh in his judgments.

Because the rapid growth of air travel had increased the risk that epidemic diseases would be imported into the United States, the EIS was becoming increasingly involved in international health. In 1957, Henderson flew down to La Plata, Argentina, to investigate an outbreak of botulism, a lethal form of food poisoning caused by a bacterial toxin. The Argentines had requested a large quantity of antitoxin, practically the entire U.S. supply at the time. Langmuir asked Henderson to accompany the shipment and make sure the antidote was used properly. After arriving in La Plata, Henderson traced the source of the botulism outbreak to a local restaurant, but by then it was too late to administer the drug.

The Argentine minister of health invited him to stay a few more days and join him for the weekend at his hunting lodge in the north of the country. But Henderson had learned of an outbreak of smallpox in Argentina and, eager to learn firsthand about an infectious disease that no longer existed in the United States, he asked to visit the affected area. The minister agreed and put him in touch with Carlos Quiros, a Peruvian epidemiologist with the World Health Organization who was helping Argentina to produce freeze-dried smallpox vaccine.

Henderson and Quiros flew in an old Pan Am clipper from Buenos Aires, following the Paraná River up north. At a regional airport in the town of Resistencia, they transferred to a Piper Cub and flew for another hour, crammed into two small seats behind the pilot. They landed at an airstrip in the pampas, close to a village where fifteen patients had been hospitalized in tents. The patients were all said to have alastrim, the mild form of smallpox that closely resembled chickenpox.

As Henderson walked down the row of beds, however, he was baffled: Some of the patients' rashes did not resemble the pictures he had seen in medical textbooks. His suspicion grew when the hospital director said that many of the patients had been admitted with smallpox and then had "relapsed" a few weeks later. Henderson knew that because a single bout with the disease produced lifelong immunity, a recurrence was impossible. He concluded that some of the patients had been admitted with smallpox and had then contracted chickenpox in the hospital, or vice versa. This was his first encounter with the complexities of the disease that would later become the central focus of his public health career.

Around this time, the world was beginning to consider a global campaign to eradicate smallpox, as Edward Jenner had proposed in the late eighteenth century. The logical entity to take up Jenner's challenge was the World Health Organization (WHO), which had been founded in 1948 as one of the specialized agencies of the United Nations and was headquartered in Geneva, Switzerland.

Several countries were already free of smallpox, having conquered the disease within their own territories through mass vaccination campaigns.

Sweden had been the first country to become smallpox-free in 1895, followed by Austria in the 1920s; England, the Soviet Union, and the Philippines in the 1930s; and the United States and Canada in the 1940s. Dr. Fred L. Soper, the director of the Pan American Sanitary Bureau, a regional office of the WHO, had launched a mass vaccination program in 1950 to eliminate smallpox from Central and South America. Although the bureau had limited resources, Soper brought in scientific advisors from North America to train the countries of the region to manufacture their own freeze-dried vaccine. Within eight years, vaccination had eliminated smallpox from the Caribbean, Central America, and South America—with the important exception of Brazil.

Despite such partial successes, however, little progress was being made in eliminating smallpox from several regions of the developing world where the disease was still endemic, including Africa, the Indian subcontinent, Southwest Asia, Indonesia, and Brazil. In 1953, the first WHO director-general, Dr. Brock Chisholm, proposed a global campaign to eradicate smallpox, but delegates from the United States and other industrialized countries rejected this initiative on the grounds that it was too ambitious, too costly, and of uncertain feasibility. Instead, the WHO decided in 1955 to launch a major program to eradicate malaria.

Meanwhile, in countries such as India, smallpox continued to cause devastating epidemics. In 1958, Albert Herrlich of the University of Munich described a severe outbreak in Bombay in which the majority of patients had "confluent" smallpox, with pustules clustered so densely that they ran together and no normal skin was visible, turning the patient's face into a hideous, oozing mask. "The head was usually covered by what appeared to be a single pustule; the nose and lips were glued together," Herrlich wrote. "When the tightly filled vesicles burst, the pus soaked through the bedsheet, became smeared on the blanket, and formed thick, yellowish scabs and crusts on the skin. . . . Swallowing was so painful that the patients refused all nourishment and, in spite of agonizing thirst, often also refused all fluids. . . . Wails and groans filled the rooms. . . . The patients were conscious to their last breath. Some . . . just lay there, dull and unresponsive. They no longer shook off the flies that sat on purulent eyelids, on the openings of mouth and nose, and in swarms on the inflamed areas of skin. But they were still alive, and with touching gestures they lifted their hands and begged for help."

In 1958, the World Health Assembly, the governing body of the WHO consisting of health ministers from all member countries, held its annual meeting in Minneapolis. This venue came about at the invitation of Minnesota Senator Hubert H. Humphrey, who was a strong supporter of the United Nations and its specialized agencies. Despite the Cold War tensions prevailing at the time, the Soviet Union decided to send a delegation to the WHO assembly after an absence of several years. From 1948 to 1957, Moscow had withdrawn its ambassadors from the United Nations and its specialized agencies. After Stalin's death in 1953, however, the Soviet hard line had begun to soften. At the Twentieth Communist Party Congress in 1956, Soviet Premier Nikita Khrushchev had made a secret speech in which he repudiated Stalinism and the inevitability of armed conflict with the capitalist West and advocated a policy of "peaceful coexistence." As an element of this new policy, Khrushchev sought to reintegrate the USSR into several international organizations, including the WHO.

At the 1958 World Health Assembly in Minneapolis, Dr. Viktor M. Zhdanov, the Soviet deputy minister of health, gave a lengthy speech in which he proposed a Soviet-style five-year plan for the worldwide eradication of smallpox. Playing off the American venue, he began by quoting Thomas Jefferson's letter to Jenner praising the invention of vaccination. Zhdanov noted that smallpox posed a threat not only to the countries where it was still endemic, but also to those free of the disease, which had to spend large sums of money to vaccinate and revaccinate their populations and to prevent infected individuals from crossing their borders. In the Soviet Union, for example, importations of smallpox from Afghanistan and Iran had spawned epidemics in southern Russia and the Central Asian republics.

Since smallpox was exclusively a human disease, Zhdanov argued that it would be easier to eradicate than mosquito-borne infections such as yellow fever or malaria, which required complex and costly efforts to control the insect carriers. Indeed, repeated attempts by the Rockefeller Foundation from 1915 to 1932 to eradicate yellow fever through mosquito control had failed miserably. Vaccinating large numbers of people against smallpox would also be more feasible than trying to treat all the patients suffering from an infection, the strategy used in earlier unsuccessful efforts to eradicate hookworm and yaws.

Zhdanov observed that the Soviet Union, through a program of compulsory vaccination, had eliminated smallpox throughout its vast and ethnically diverse territory by 1936. This program had been conducted at a time when the country's transportation networks and health services were rudimentary, the available vaccine was of poor quality, and the Central Asian republics were no better off economically than most developing countries. If smallpox had been eliminated from the Soviet Union under such difficult conditions, why could it not be eradicated globally? Zhdanov proposed that WHO member countries devote two years to producing large amounts of freeze-dried vaccine, followed by a three-year program to vaccinate 80 percent of the population in endemic areas, the level believed sufficient to halt smallpox transmission. To help kick off the campaign, the Soviet Union pledged to donate 25 million doses of vaccine.

Despite Zhdanov's impressive vision, the Soviet proposal aroused little enthusiasm. The WHO was already deeply committed to the malaria eradication campaign, which the United States was supporting extravagantly. Although the malaria program had a staff of more than five hundred people and an annual budget of $13 million, it was making little headway. For an organization the size of the WHO, whose entire budget was just over $30 million, launching a second eradication campaign would place a major drain on resources. Moreover, the difficulties facing the malaria program, combined with the earlier failed attempts to eradicate yellow fever, hookworm, and yaws, meant that such initiatives were viewed with considerable skepticism.

The WHO director-general, Marcolino Candau of Brazil, had received his public health training in the ill-fated yellow fever program and doubted the technical feasibility of eliminating smallpox, which was still infecting millions of people. In his view, a successful eradication campaign would require vaccinating every human being on the planet, yet that objective was clearly impossible because primitive tribes living in remote areas such as the Amazon rain forest had not been seen by outsiders in decades. How could one expect to conquer smallpox without finding all the unprotected people and vaccinating them?

Despite widespread skepticism about the feasibility of eradication, the WHO member countries wanted to be cordial to the Soviets as they returned to the organization. As a result, the next World Health Assembly in 1959 unanimously approved the Soviet proposal after little debate. The amount of money allocated for smallpox eradication was minuscule, however: only about $300,000 per year. No mention was made of the vast discrepancy between the program's ambitious goals and its tiny budget, nor were member states asked to provide voluntary contributions.

To implement the 1959 decision, the WHO appointed a medical officer and a secretary, who for several years were the only full-time employees at the Geneva headquarters working on smallpox eradication. The campaign had no dedicated budget, no deadline, and no organizational structure or management control. Between 1962 and 1965, the WHO launched modest vaccination programs in five small countries, but these efforts were hobbled by a lack of money, vehicles, and manpower. Most of the smallpox vaccine was in liquid form and, if not continuously refrigerated, lost its potency within forty-eight hours. Although the Soviets donated limited amounts of freeze-dried vaccine and some of the endemic countries learned to produce their own, the quality and potency of the vaccine were inconsistent.

Meanwhile, D. A. Henderson was continuing his training in the field of public health. After he completed his tour with the Epidemic Intelligence Service in 1957, he and Alex Langmuir discussed how to retain EIS officers in the U.S. Public Health Service after their two years of compulsory military service were over. With this goal in mind, the two men developed a five-year post-EIS training program that entailed two years of medical residency, one year of additional study, and two years with the Public Health Service to provide a solid foundation for a government career. After the five-year training program had been approved, Henderson decided to sign up. He began his residency in internal medicine at the same hospital in Cooperstown, New York, where he had done his internship.

Toward the end of his two-year residency, Henderson began to have serious doubts about devoting his career to combating infectious diseases. Now that powerful antibiotics and vaccines had banished most of the old plagues from the industrialized world, the field had become a professional backwater and morale was low. A more promising future seemed to lie in the study of chronic diseases, such as cancer and heart disease. Accordingly, Henderson enrolled in a one-year master's program in chronic disease epidemiology at the Johns Hopkins University School of Hygiene and Public Health, completing his degree in 1960. By that time, however, he had lost interest in longitudinal studies of disease causation, which involved devising elaborate survey questionnaires and took several years to complete. He simply did not have the patience for that kind of research.

Instead, Henderson returned to the CDC in Atlanta to become assistant chief of the epidemiology branch, reporting directly to Dr. Langmuir. At that time, the Public Health Service was focusing primarily on domestic outbreaks of infectious disease, responding to foreign epidemics only when they directly threatened the United States. But Henderson believed that it would be easier for the CDC to retain good people if they could get involved in international health.

One activity in particular that made sense was to be better prepared for importations of smallpox. Although the disease had been eliminated from the United States in the mid-1940s, the Public Health Service still spent large sums of money to vaccinate the U.S. population and to keep active small-pox cases from entering the country. Preventive measures included post-ing health officers at ports and airports to quarantine people with suspicious rashes and requiring travelers to carry smallpox vaccination certificates. Nevertheless, if a contagious case of smallpox were to slip through the net and trigger a major epidemic, the U.S. government would not be well equipped to handle it. In 1947, a small outbreak of smallpox in New York City had triggered widespread panic, forcing city health officials to launch a chaotic monthlong campaign in which 6.3 million people were vaccinated. Ironically, six people had died from adverse reactions to the vaccine—three times the number of deaths caused by the disease itself.

In an effort to improve the ability to respond to importations, Henderson created a Smallpox Surveillance Unit at the CDC and staffed it with four officers. Each time an importation of smallpox occurred in Western Europe, he dispatched a member of the team to observe how the outbreak was contained and what problems arose. In the winter of 1961–62, for example, sixty cases of smallpox in England and Wales were traced to trav-elers from Pakistan. To contain the outbreak, 5.5 million people were vac-cinated, and complications from the vaccine caused at least eighteen deaths. Also during the 1960s, West Germany built two special isolation hospitals that were kept fully equipped for use only when smallpox cases were im-ported into the country.

The Smallpox Surveillance Unit responded to every suspect case of smallpox in a visitor entering the United States. In 1962, a fourteen-year-old Canadian boy traveled by plane from Brazil to New York City and then by train to Toronto. Shortly after arriving home, he was hospitalized with alastrim, the mild form of smallpox, raising fears that he could have exposed a large number of Americans en route. Despite intensive surveillance, however, no further cases were detected. In another scare in May 1965, a West African woman flew from Ghana to Washington, D.C., where she developed a skin rash that was initially diagnosed as smallpox but later turned out to be chickenpox. Although both incidents were false alarms, they demonstrated that international air travel had greatly increased the risk of importations.

The CDC also undertook to assess the frequency of serious complications from smallpox vaccination. People whose immune system was suppressed, such as children with acute leukemia or organ transplant recipients, often failed to keep the replication of the vaccine virus in check and developed a condition called "progressive vaccinia," in which the primary vaccination pustule failed to heal and secondary skin lesions appeared all over the body. The lesions continued to expand grotesquely for two to five months until, in most cases, the patient died. In pregnant women who were vaccinated, a condition known as "fetal vaccinia" could cause the fetus to be stillborn or to die a few days after birth. And in individuals suffering from eczema, the vaccine virus could replicate out of control in the affected areas of skin, causing an intense inflammation known as "eczema vaccinatum." This complication could be fatal in infants if large areas of skin were affected.

Other complications of smallpox vaccination were known to occur in otherwise healthy people. In a few rare instances, the recipient developed an extensive rash six to nine days after vaccination, sometimes covering the entire body. Known as "generalized vaccinia," this condition generally had a good prognosis. In a fraction of normal people, however, primary smallpox vaccination produced an inflammation of the brain known as "postvaccinal encephalitis," which was fatal in 25 percent of cases and could cause permanent brain damage.

In 1965, Dr. Henry Kempe, a pediatrician in Denver, Colorado, advocated halting the routine vaccination of U.S. children against smallpox on the grounds that the risks outweighed the benefits. He pointed out that since 1948, between two hundred and three hundred children had died from complications of smallpox vaccination, yet only a single child had died from the disease itself. Kempe's critics argued that he was exaggerating the risks of the vaccine, but reliable data on the frequency of serious complications did not exist.

In an attempt to resolve the controversy, the CDC team reviewed government statistics and conducted a national survey of physicians to determine how many adverse reactions had been reported. They found that of 5.6 million primary smallpox vaccinations and 8.6 million revaccinations each year, complications from the vaccine caused about nine deaths and twelve nonfatal cases of post-vaccinal encephalitis, with permanent damage in 30 percent to 40 percent of cases. Because the risk of complications was relatively high in babies, the CDC recommended that primary smallpox vaccination be delayed until after the first year of life. Mandatory vaccination before school entry remained in effect, however, because of the need to maintain a high level of "herd immunity" in the general population, thereby preventing possible importations of smallpox from triggering massive epidemics.

Some complications of smallpox vaccination could be treated by injecting the affected individuals with a preparation of antibodies against vaccinia virus known as vaccinia immune globulin, or VIG. This product was derived from the blood plasma of volunteers who had been vaccinated repeatedly against smallpox. Nevertheless, VIG was of no value in treating post-vaccinal encephalitis, and no reliable ways were found of screening potential vaccinees in advance to predict who would develop this serious complication.

Another task undertaken by the CDC Smallpox Surveillance Unit was to evaluate new ways of administering the smallpox vaccine. The traditional vaccination method in the United States and Europe was to place a drop of vaccine on the recipient's arm and scratch it into the skin with a needle, but this technique was time-consuming and did not always yield good immunity. During World War II, the U.S. Army had developed an electric-powered jet injector gun for immunizing large numbers of soldiers that was rapid, effective, and relatively painless. In 1962, Henderson met with the inventor of the jet injector, Aaron Ismach, who proposed to modify the nozzle so that it would deliver smallpox vaccine as a concentrated spray into the outer layers of the skin. The army contracted with the CDC to test the new device in a population of sixty thousand people on the Polynesian island of Tonga. This study demonstrated that under optimal conditions, the jet injector could successfully vaccinate one thousand people an hour.

Nevertheless, injector guns had serious drawbacks for field operations in developing countries because they relied on electrical generators, which frequently malfunctioned or went out of service. To address the problem, Ismach developed a mechanical version of the jet injector that was powered by a hydraulic pump operated by a foot pedal. This "ped-o-jet" system significantly reduced the need for maintenance and spare parts. The CDC Smallpox Surveillance Unit did a field study in Brazil to assess the cost of the ped-o-jet compared with standard vaccination techniques, and obtained promising results.

Meanwhile, the WHO smallpox eradication campaign was making little headway, and the Soviets were becoming increasingly impatient with the program's slow progress and meager funding. The United States, for its part, remained deeply committed to malaria eradication and opposed the creation of a competing effort—a schism echoing the deeper ideological and geopolitical conflicts of the Cold War. In a world dominated by two hostile superpowers, it would take the fortuitous emergence of cooperation between Washington and Moscow to move smallpox eradication to the front burner.

The change in U.S. policy came about in an unexpected way. In 1962, the Merck pharmaceutical company developed a new measles vaccine and arranged to have the U.S. National Institutes of Health test its effectiveness in Upper Volta, a West African country where measles was a major killer of young children under age six. During the summer of 1963, field workers used jet injectors to deliver measles vaccine to more than ten thousand children, with no significant adverse effects. The Voltaic minister of health, Paul Lambin, was enthusiastic about the measles vaccination program and persuaded the United States to expand it to more than 700,000 Voltaic children, causing a sharp drop in the incidence of the disease.

On learning of the success of measles vaccination in Upper Volta, other West African countries asked the Agency for International Development (AID), the foreign-aid arm of the U.S. government, to include them in the campaign. In late 1963, AID agreed to establish measles vaccination programs in six additional francophone countries, providing vehicles, jet injectors, and vaccine to immunize children between the ages of six months and six years. In addition to the humanitarian rationale, the U.S.-funded campaign promised to increase American political and economic influence in francophone West Africa, a region historically dominated by France.

AID asked the CDC to assign an officer who would monitor the vaccination programs and send back periodic reports. In December 1964, Henderson sent a young physician named Lawrence Altman, who later became a medical reporter for the *New York Times*. Altman was supposed to visit West Africa for six weeks but ended up staying more than six months because the AID vaccination program was in such disarray. The large trucks used to distribute the measles vaccine got only four miles per gallon, frequently became stuck in the sand, and had small butane refrigerators that did not keep the vaccine cold, so it deteriorated rapidly.

Undaunted by the poor performance of the measles vaccination program, AID planned to expand it from six to nine countries of francophone West Africa in 1965–66. This time, the aid agency asked the CDC to station one officer in each country for six months to assist with the vaccination effort, but Henderson could not spare nine people from his staff of forty. He also believed that the measles control program was fundamentally ill-conceived. Because babies were continually being born, the disease would inevitably return unless the vaccination campaign was repeated every year. Yet AID planned to vaccinate 25 percent of the children in the participating countries during each of the next four years and then pull out. Since the measles vaccine cost about a dollar per dose, none of the West African countries could afford to keep vaccinating after the program ended. Henderson worried that if the AID-funded vaccination campaign stopped abruptly, renewed outbreaks of measles in young children would spark public anger and even social unrest.

The U.S. government was already committed to the measles vaccination program in West Africa, so canceling it was not an option. Henderson also hoped to expand the CDC's involvement in international health activities and did not wish to alienate AID by declining to participate in the measles program. As a diplomatic way out, he decided to make the aid agency a proposal he was certain they would refuse: combining the measles control effort with a new program to eliminate smallpox from the entire West African region.

Henderson believed that this ambitious goal was technically feasible. Smallpox vaccine could be administered efficiently to large numbers of people with the ped-o-jet system, and because the vaccine cost only about a penny per dose, the program would be financially sustainable over the long haul. Moreover, whereas the benefits of measles control were temporary, once smallpox had been eliminated from West Africa, the disease would be unlikely to return. A successful effort in this heavily endemic region would also provide an important stimulus to the WHO's global eradication campaign.

Henderson's proposal, drafted in the summer of 1965, called for a five-year program of smallpox elimination and measles control in eighteen contiguous countries of West and Central Africa, including English-speaking Nigeria, Ghana, Liberia, and Sierra Leone. Under the plan, AID would supply the vaccine, vehicles, and equipment, while the CDC would provide technical assistance. Henderson conditioned the CDC's participation, however, on coverage of the entire West African region. The rationale was that if some of the contiguous countries were left out, smallpox could be easily reintroduced from the excluded areas. This fact was particularly true of Nigeria, which lay at the crossroads of the region.

As Henderson had expected, AID officials accepted the idea of a combined measles/smallpox vaccination campaign but balked at the inclusion of Nigeria, which accounted for more than half the total population of the eighteen countries. Not only would including Nigeria double the overall size and cost of the effort, but AID opposed a new health program in that country for political and economic reasons. The United States was already providing substantial financial assistance to Nigeria and, according to the "trickle-down" theory popular at the time, economic development would inevitably result in improved health services and conditions. For this reason, a dedicated health program in Nigeria was the last thing that AID wanted. Unwilling to accept Henderson's conditions but aware that it would need to rely on the CDC for technical assistance, the aid agency put off a final decision.

Meanwhile, the United Nations had named 1966 as International Cooperation Year to commemorate the twentieth anniversary of its first session in 1946. President Lyndon B. Johnson wanted to announce an appropriate U.S. initiative to mark the occasion, and in November 1965, the White House staff called a meeting of interested government agencies to discuss various options. During this meeting, the Public Health Service representative presented the CDC's proposal for a combined program of smallpox elimination and measles control in West Africa. Members of the White House staff decided that the expanded vaccination campaign would be an ideal initiative for International Cooperation Year, and President Johnson agreed.

The president's decision to endorse the CDC proposal caught everyone by surprise. AID was irate, Henderson was dumbfounded, and Langmuir was horrified by this unexpected turn of events. Because the primary mission of the epidemiology branch was training EIS officers, Langmuir feared that assuming operational responsibility for a major international health campaign would divert attention and resources from the educational program that he had worked so hard to establish. Henderson cringed before the wrath of his boss and mentor. "Alex, this was not my intention," he pleaded, explaining that he had developed a proposal that made sense from a public health perspective, secure in the knowledge that AID would turn it down. But President Johnson's unexpected intervention had foiled his clever scheme.

When Henderson suggested that the epidemiologists recruited for the West African campaign might be integrated with the Epidemic Intelligence Service, Langmuir angrily rejected the idea. "Take whomever you want and set up the program outside the epidemiology branch," he said bitterly. "I don't want you here." After that exchange, he and Henderson did not speak for the next two years.

Before the West African program could begin, Henderson and his staff had to negotiate agreements with each of the participating countries, establish good working relations with AID and the WHO, and recruit, train, and move to Africa a large team of epidemiologists and their families. AID officials anticipated that it would take three years to set up a program of this magnitude, but Henderson aimed to complete all the preparations within the next fifteen months.

Around this time, the international community began to express support for an intensified global campaign to eradicate smallpox. The Soviet Union had long pressed for a major expansion of the seriously underfunded WHO program. At the World Health Assembly in 1965, for example, the Soviet delegate complained that malaria eradication was "the favorite daughter of WHO, whereas smallpox eradication seemed to have been treated rather as a foster child." By the mid-1960s, the WHO had made little headway against smallpox in the major endemic zones of Africa, Brazil, Indonesia, and South Asia. Citizens of nearly fifty countries still suffered from the disease on an ongoing basis or were at high risk of importations from neighboring areas. According to a history of the program by June Goodfield, "Poor countries just could not undertake campaigns; others found their efforts thwarted as war and the movements of refugees carried smallpox from place to place; sufficient vaccine did not exist; nor was it always effective. As their programs faltered, the endemic countries became disheartened and even in those that claimed successful mass vaccination, smallpox was flaring again."

Scientific skeptics continued to argue that the very idea of disease eradication was misguided. The famous French bacteriologist René Dubos devoted an entire chapter of his 1965 book, *Man Adapting,* to an explanation of why the eradication—as opposed to control—of infectious diseases was scientifically impossible. Eradication programs, he concluded, "will eventually become a curiosity item on library shelves, just as have all social utopias."

Despite the severe limitations of the WHO smallpox eradication campaign, twenty-four countries became free of the disease during the 1960s, mostly through their own efforts. The most notable case was the People's Republic of China. After the Communist Party assumed power in 1949, it constructed factories for the large-scale production of liquid smallpox vaccine and launched a mass vaccination campaign in October 1951. Fourteen years later, in 1965, smallpox had been eliminated from the entire country. Had the WHO learned of this success at the time, it would have provided a major source of optimism, but substantive information to confirm the elimination of smallpox from China did not become available until 1979.

By early 1966, the United States was warming to the idea of intensifying the smallpox eradication program, in part because of the serious problems facing the malaria campaign. At the Nineteenth World Health Assembly in May 1966, however, the U.S. delegation was deeply split. Now that the AID/CDC smallpox vaccination program in West Africa was getting organized, the U.S. Public Health Service favored devoting more WHO resources to smallpox eradication. But the State Department countered that the budgets of the specialized UN agencies were growing too fast and that it was necessary to put a damper on spending.

WHO Director-General Candau deeply resented the fact that the United States was asking the organization to intensify its efforts to eradicate smallpox but was unwilling to provide additional funds for that purpose. In a gambit to torpedo the proposed resolution, Candau called for a supplementary allocation of $2.4 million for smallpox eradication on top of the WHO regular budget. He declared that if the member countries did not approve the additional money, it would be impossible to launch the new initiative. Having established this condition, the director-general believed that the United States and other wealthy countries would reject an intensified smallpox program because they would have to foot the lion's share of the bill.

Candau had miscalculated, however. The industrialized countries had a direct stake in smallpox eradication because they were already investing a great deal of money to vaccinate their populations and to prevent importations of smallpox from countries where the disease was still endemic. In the United States alone, smallpox prevention efforts cost more than $150 million a year. For this reason, Washington finally decided to make common cause with Moscow and get serious about eradicating the disease. After a heated debate, the World Health Assembly took a formal vote—a rare event in a forum where decisions were customarily made by consensus. The smallpox resolution passed by two votes, the closest margin in the history of the WHO. It is not known which countries voted against the resolution because the delegates simply raised their hands and no official roll call was taken.

The decision by the World Health Assembly established a line item in the regular WHO budget to fund an Intensified Smallpox Eradication Program, which would seek to eliminate the disease from the planet within ten years. This deadline echoed President John F. Kennedy's famous challenge, in a speech to Congress on May 25, 1961, that the United States "should commit itself to achieving the goal, before this decade is out, of landing a man on the moon and returning him safely to the earth." But instead of a race between the superpowers driven by Cold War rivalry, the intensified smallpox eradication campaign was to be a cooperative effort involving some fifty countries around the world.

Director-General Candau was furious over this unexpected development. He called U.S. Surgeon-General Luther L. Terry, the director of the Public Health Service, and complained bitterly that the new smallpox eradication campaign was certain to fail and would discredit the WHO. For this reason, he insisted on putting an American medical officer in charge of the program so that the United States would bear most of the blame for its inevitable collapse. Candau even had someone in mind for the job.

In June 1966, the surgeon-general's office told D. A. Henderson to travel from Atlanta to Washington for urgent consultations. When Henderson arrived at Public Health Service headquarters, James Watt, the director of the Office of International Health, broke the news that he had been chosen to head the intensified WHO campaign to eradicate smallpox. Stunned, Henderson tried to turn down the assignment. He was reluctant to leave the CDC now that the West African program, of which he was the primary architect, was just getting under way. Eliminating smallpox from a major region of the world was surely enough of a challenge, and he felt a responsibility to the people he had hired. Henderson was also convinced that the WHO was a dysfunctional bureaucracy that was incapable of managing a global disease eradication campaign—particularly given the director-general's obvious lack of commitment to that goal.

Dr. Watt replied that Candau had specifically requested him by name, adding, "You're ordered to go, in this case."

Henderson was taken aback. The usual practice in the Public Health Service was to discuss career opportunities, not to order people to do things. "Suppose I refuse?" he asked.

"Well, you could always resign," Watt replied.

"Good heavens, are you serious?"

"Yes, we are."

They finally worked out a compromise by which Henderson would go to Geneva for eighteen months to set up the program and would then return to the CDC. If, at any point during his stay, he found his position untenable, he could send a cable to Washington requesting another assignment. Believing that his tour of duty at WHO headquarters would be short, Henderson put half of his furniture and household goods in storage in Atlanta and moved to Geneva in November 1966 with his wife, three children, and two cats. As it happened, he did not see his stored furniture again for eleven years.

FIVE

GLOBAL ERADICATION

When D. A. Henderson arrived at WHO headquarters in Geneva in November 1966, he was an energetic and ambitious man of thirty-eight, eager to take on the greatest challenge of his career. Little remained of the smallpox eradication program that the WHO had launched in 1959 at the request of the Soviet Union. The former head of the program, an Egyptian, had been fired several months earlier for gambling, which had taken him away from his duties for long periods. By the time Henderson took over, the only WHO staff members still assigned to the program were Isao Arita and a secretary. Arita, a low-key Japanese virologist, had worked for two years on smallpox eradication in Liberia. Although he had made little headway against the disease, he had profited from the experience by learning reasonably good English.

Henderson set up a new Smallpox Eradication Unit within the Division of Viral Diseases on the sixth floor of the eight-story WHO headquarters building. From the outset, the program faced daunting challenges. In 1967, smallpox was still entrenched in five parts of the world: Brazil, West and

Central Africa, East and Southern Africa, the Indian subcontinent, and Indonesia. Within these regions, thirty-one countries were infected on a continual basis, and twelve others experienced importations that year. The total number of smallpox cases was estimated at 10 million to 15 million annually, with 2 million deaths. In fact, the true magnitude of the problem was unknown: Widespread nonreporting and underreporting of cases had created a huge gap between the official statistics and reality.

One of Henderson's first tests as program director came during the World Health Assembly in May 1967, when he arranged to meet with the Soviet delegate, Dr. Dimitry Venediktov, the USSR's deputy minister of health. Henderson was anxious about the meeting because he had heard that Moscow had protested his appointment. The Soviets had naturally assumed that the WHO would reward their persistent support of small-pox eradication and their generous donation of vaccine by selecting a Russian to head the program. Instead, much to their dismay, an American had been chosen. Henderson feared that if the Soviets were sufficiently dis-gruntled, they might back out of their commitment to supply 25 million doses of smallpox vaccine each year. Since the USSR had a vaccine produc-tion capacity far exceeding that of any other country, it was unclear who would or could take the Soviets' place.

During their meeting, however, Venediktov was surprisingly friendly. He told Henderson that the Soviet investigative services had checked him out and determined that he was a serious scientist and not a political op-erative. The deputy minister then pledged Moscow's full support for the intensified smallpox eradication program, to include the annual donations of vaccine. Not to worry, he joked: Once the centrally planned Soviet economy geared up to produce a large quantity of anything, it was nearly impossible to turn off the tap.

Despite the welcome support from Moscow, Henderson encountered widespread skepticism from other quarters about the feasibility of small-pox eradication. Part of the problem was that the earlier campaigns to eradicate hookworm, yaws, yellow fever, and malaria had all failed—the last after an investment of more than $2 billion. The United Nations Children's Fund (UNICEF), which had been an active participant in the malaria program, announced during the 1967 World Health Assembly that it had lost confidence in disease eradication and would contribute nothing to the smallpox campaign. Needless to say, this declaration dampened the enthusiasm of other potential donors.

Still, Henderson remained optimistic about the ultimate prospects for success because smallpox had several characteristics that made it a particularly good target for global eradication. As a contagious disease that killed a third of its victims, smallpox was in a class by itself as a public health threat, giving governments a strong political incentive to vaccinate their populations. Diagnosis and surveillance of smallpox were also easier than with other infectious diseases. In contrast to the causative agents of yellow fever and malaria, variola virus had no animal or insect carrier that harbored it when human hosts were unavailable. The mosquitoes that transmitted malaria had foiled attempts at environmental control by evolving resistance to the pesticide DDT. But because smallpox infected only humans, once the disease had been eliminated from the world's population, it would be unable to return.

Smallpox had other features that promised to ease the task of eradication. Unlike serum hepatitis virus, which could infect a person for years, variola virus always produced an acute illness that progressed rapidly, culminating within a month either in death or in recovery with permanent immunity. Whereas polio virus often produced silent or atypical infections, smallpox patients always developed a distinctive facial rash, making it possible for health workers to identify cases easily and to control the spread of infection. In contrast to leprosy and venereal diseases, smallpox was not associated with a cultural stigma and hence was unlikely to be deliberately concealed. And unlike influenza virus, which rapidly shuffled its surface proteins to evade human immune defenses, the antigens of variola virus were so stable that a single bout of smallpox induced lifelong immunity, and the same strain of vaccinia could be used to vaccinate year after year.

Last, but not least, a freeze-dried smallpox vaccine was available that was easy to manufacture, cost only about a penny a dose, protected for several years with a single inoculation, and was relatively stable in warm climates, reducing the need for refrigeration. Whereas most vaccines took months to induce immunity, the smallpox vaccine acted with remarkable speed, providing nearly total protection within ten to twelve days. As a result, vaccinating people who had been exposed to the virus but had not yet developed symptoms, even three to five days after infection, often prevented smallpox entirely and usually avoided a fatal outcome. Moreover, vaccination of people who were at risk for smallpox but had not yet been exposed could serve to prevent transmission, containing the spread of an epidemic. All these factors made the eradication of smallpox a theoretical possibility. The real challenge was to develop the organizational, managerial, and logistical structures needed to carry out vaccination campaigns on a large scale.

As Henderson had feared, he soon came into conflict with the hidebound WHO bureaucracy. The budget for smallpox eradication was allocated directly to four semiautonomous regional WHO offices that were responsible for the parts of the world where the disease was still endemic: the Americas, Southeast Asia, the Eastern Mediterranean, and Africa. Not only did the regional directors resist taking orders from WHO headquarters, but they also hesitated to intervene in the internal health affairs of member states. As a result, the WHO offices for Africa and Southeast Asia—the two regions most seriously affected by smallpox—did not promote the smallpox eradication campaign and provided assistance only to those countries specifically requesting it. Director-General Candau, who was deeply skeptical about the feasibility of smallpox eradication, did not attempt to fix this dysfunctional arrangement. In the face of such daunting bureaucratic obstacles, Henderson and his team had to bend long-standing rules and regulations to get the job done. Not surprisingly, their freewheeling management style aroused resentment from senior WHO officials.

In planning strategy for the smallpox eradication campaign, it was essential to prioritize. The AID/CDC program in West Africa, which got under way in January 1967, was staffed and funded by the U.S. government as a bilateral operation, with the WHO providing oversight and modest financial assistance. Henderson was happy to let the CDC take the lead for this major region, which had a population of more than 100 million people and an extremely poor infrastructure of roads, telecommunications, and health facilities. Meanwhile, the Smallpox Eradication Unit decided to focus its initial efforts on Brazil and Indonesia for geographic reasons. Brazil was the only country in South America where smallpox was still endemic, whereas Indonesia was a vast archipelago with no contiguous neighbors. Once smallpox had been eliminated from these countries, it would be possible to keep them free of the disease with minimum effort. The WHO would then be in a good position to shift resources elsewhere.

Although the government of each smallpox-endemic country was nominally responsible for directing its own vaccination campaign, WHO staff epidemiologists served in an advisory capacity, providing technical and administrative support and training local health workers, who were often poorly educated and undisciplined. Between two and four expatriate staff members were assigned to a given country, and the total number of international advisers deployed worldwide never exceeded one hundred at any given time.

Henderson understood that the effectiveness of the smallpox eradication program would depend, to a large extent, on the quality of the expatriate staff members stationed in each country. When recruiting personnel, he decided not to require any prior experience of working with smallpox. Instead, he sought out people with basic intelligence, good problem-solving skills, and the motivation needed to work long hours in the field under harsh living conditions. As a result, the international advisers working on the smallpox eradication campaign were young, energetic, and idealistic. Nearly all were under forty and many were under thirty. Unconventional in dress and appearance, they often sported beards and long hair that made them stand out awkwardly at embassy cocktail parties. The most effective ones leavened their youthful idealism with common sense, resourcefulness, and stubborn persistence.

Henderson believed that only by spending a great deal of time in the field could the international advisers learn what was really going on and build strong working relationships with local health workers. As a result, he decreed that all WHO staff members assigned to the Smallpox Eradication Unit, including those based at WHO headquarters in Geneva, would have to spend at least a third of their time working in rural areas, away from the relative comfort of the capital cities. Those who failed to comply with this rule came under a fair amount of pressure and usually ended up requesting another assignment.

Although the Cold War was at its height, Soviet and American members of the WHO international staff worked together closely and professionally, albeit with a certain amount of intrigue. A few Soviet advisors were obviously gathering information for their superiors in Moscow, although none was ever exposed as a spy. One Russian who appeared to be working as a double agent had too many responsibilities to do an effective job in the field and was dismissed for that reason. Ironically, most of the tension was not between Soviets and Americans, but among the Soviets themselves. Whenever Henderson invited Soviet staff members to his home for dinner during the early years of the program, his guests were always accompanied by an uninvited KGB escort.

It was obvious that unless all the countries where smallpox was endemic participated constructively in the WHO campaign, achieving the goal of global eradication would be impossible. But Henderson encountered varying degrees of difficulty in securing the cooperation of the national health authorities in the affected countries. Some did not consider smallpox a high

priority and had no intention of taking aggressive action, while others refused even to admit that the disease was present. Moreover, the ministries
of health were always desperately short of funds and were generally unwilling to invest their meager resources on smallpox eradication. For these
reasons, Henderson had to cajole several of the endemic countries into
launching vaccination campaigns. The fact that the WHO could offer free
vaccine, vehicles, and fuel—all valuable resources in poor countries—was
the carrot that got many of the programs started.

As the campaign progressed, the Smallpox Eradication Unit tried to exert
political pressure on governments to cooperate, although its sources of
leverage were limited. One useful tool was the publication of progress
reports in the *Weekly Epidemiological Record,* an official WHO journal.
These reports surveyed the smallpox eradication programs in various countries, noting the ones that were most successful. Since recalcitrant governments usually did not want to receive a bad review from the WHO, they
thus had some incentive to work harder.

A second tool was the annual status report on the smallpox eradication
program presented to the World Health Assembly each May. Director-
General Candau was not eager to highlight the smallpox program because
he continued to question its ultimate prospects for success. But the Soviet
Union or one of its Eastern Bloc allies would always insist that smallpox
eradication be placed on the agenda, and Henderson discreetly encouraged
these efforts. Once it was widely known that the smallpox issue would be
addressed, national health ministers were put on notice that they would
have to account for their accomplishments—or the lack thereof. Because
having nothing positive to say would be embarrassing, governments often
launched new initiatives just before the annual meeting.

The Soviet Union never wavered in its political and material support
for the WHO smallpox eradication program, donating large amounts of
vaccine and offering the services of several skilled epidemiologists and a
diagnostic laboratory. Even during the chilliest years of the Cold War,

smallpox eradication was one of the few areas in which tacit cooperation between the superpowers continued. At the World Health Assembly in 1971, for example, the Soviet delegate gave a speech in which he angrily condemned U.S. aggression in Vietnam and then, without skipping a beat, offered to increase the Soviet contribution of smallpox vaccine. Washington and Moscow also pressured their respective client states to cooperate with the WHO eradication campaign.

In addition to the political and bureaucratic challenges, two key technological innovations set the stage for the conquest of smallpox: a heat-stable vaccine and a simple but effective delivery method. Throughout the 1950s, most countries had relied on liquid smallpox vaccine that rapidly lost its potency without refrigeration, making it necessary to establish a "cold chain" to keep the vaccine chilled throughout its journey from the production plant to the recipient's arm. This requirement had created major logistical problems in tropical developing countries. Accordingly, one of Henderson's top priorities was to arrange for the large-scale production of freeze-dried vaccine, which was much less dependent on refrigeration.

Freeze-dried smallpox vaccine was manufactured by inoculating vaccinia virus into scratches made on the flanks and bellies of living calves. Pustules appeared on the animals' skin and were harvested after seven to nine days by scraping them with a sharpened stainless-steel spoon. The virus-rich pus, known as "vaccine pulp," was inevitably contaminated with animal hair, skin cells, bacteria, and other viruses. This material was filtered to remove the coarser particles, a stabilizer was added, and the suspension was poured into small glass vials. Freezing under a vacuum caused the liquid to evaporate, leaving a dry pellet at the bottom of each vial, which was then vacuum-sealed. Prior to use, the vaccine was rehydrated by diluting it with a solution of glycerine and saline. Although the WHO recommended that freeze-dried smallpox vaccine not be kept out of the refrigerator for more than a month, in practice it remained potent at room temperature for much longer. Vaccinators often carried a few vials of freeze-dried vaccine in a shirt pocket when they went into the field.

Ensuring an adequate supply of vaccine was the most critical task facing the Smallpox Eradication Unit at the start of the intensified program. More than 30 million doses per year (not including the doses that the United States supplied to the CDC-led West African country campaigns) were required for countries that were not manufacturing their own vaccine or were producing insufficient quantities. Since the program budget was too limited to cover the cost of vaccine procurement, member governments were asked to donate vaccine to the smallpox-endemic countries. The Soviet Union provided nearly two-thirds of the vaccine channeled through the WHO, and twenty-six other countries contributed smaller amounts. Consignments of vaccine were shipped to Geneva and kept in a large refrigerated storage locker that the WHO rented from a fruit warehouse. Whenever a country program office called with a request for vaccine, the Smallpox Eradication Unit could arrange for delivery by airplane anywhere in the world within seventy-two hours.

The quality of the smallpox vaccine used in the campaign posed another challenge. When the WHO began testing lots of donated vaccine in 1967, the quality was surprisingly poor. Of thirty-five batches, twenty-two had to be rejected because they failed to meet basic standards of purity, potency, and stability. This finding indicated the need for regular testing of the vaccine used in the eradication program, whether donated or manufactured by the endemic countries themselves. Although Henderson was informed that the WHO did not have the authority to impose a worldwide quality standard for smallpox vaccine, he was not deterred by the lack of precedent. Instead, he helped to draft a resolution, later passed by the World Health Assembly, requiring all the vaccine used in the eradication campaign to meet agreed-upon standards of safety and effectiveness.

Early on, testing revealed that some batches of Soviet smallpox vaccine were seriously deficient: Not only was the potency substandard, but the level of bacterial contamination was unacceptably high. Because the USSR was providing the lion's share of vaccine for the WHO eradication program, Henderson decided to travel to Moscow for urgent consultations with Soviet health officials about the problem. When the WHO deputy director-general learned of the planned trip, he tried to cancel it because of political sensitivities about irritating the Soviets. Henderson smoothed bureaucratic feathers at headquarters by promising to discuss other issues. But soon after his arrival in Moscow, he sat down with Dimitry Venediktov and other senior health officials and raised the subject of the deficiencies in the Soviet vaccine.

After the Soviets had studied the data, they thanked Henderson for the information and promised that all future batches of smallpox vaccine would meet the WHO's standards of potency and quality. Venediktov later reported that he had shut down several substandard production plants and established a centralized testing laboratory. Although the WHO required a minimum concentration of 100 million particles of vaccinia virus per milliliter, the improved Soviet vaccine was about ten times more concentrated. This higher potency meant that the vaccine remained effective in the field even after some deterioration from exposure to heat.

Another important goal of the smallpox eradication program was to promote vaccine production in the endemic countries themselves so that they would gradually become self-sufficient. With assistance from WHO consultants, several countries established indigenous laboratories to manufacture freeze-dried smallpox vaccine and to perform their own quality testing. Five years into the program, 80 percent of the vaccine in use was being produced in the endemic countries, and some of these countries even supplied vaccine to others.

The WHO also made an effort to standardize the strains of vaccinia virus used in the smallpox vaccine. In 1968, seventy-one vaccine manufacturers around the world were employing more than nineteen different strains of vaccinia. To encourage standardization, the WHO promoted the use of two strains that were known to cause a lower frequency of complications: one developed in 1876 by the New York City Board of Health, and the other in 1916 by the Lister Institute in England. By 1972, two-thirds of the smallpox vaccine manufacturers around the world were using the Lister strain. The WHO also established a reference laboratory for smallpox vaccine in Bilthoven, the Netherlands, that shipped seed cultures of the Lister strain to vaccine producers on request.

Beyond an ample supply of high-quality vaccine, the smallpox eradication campaign required a cost-effective means of delivery. Whereas most vaccines had to be injected beneath the skin or into muscle to induce immunity, the smallpox vaccine was most effective when introduced into the superficial layers of the skin, where the replication of vaccinia virus stimulated the recipient's immune system. When someone was vaccinated for the first time, a pustule formed at the vaccination site, indicating a successful "take." The pustule then scabbed over, leaving a permanent scar. Revaccination caused a smaller reaction that usually did not leave a scar.

Before the late 1960s, the standard method of smallpox vaccination was to place a drop of vaccine on the patient's arm and press it into the skin with a single-point needle, five times for a primary vaccination and fifteen times for a revaccination. This method was wasteful of vaccine and difficult for health workers to perform correctly. In many cases, the vaccinator did not press hard enough, and an insufficient amount of vaccine entered the skin to induce a protective level of immunity.

The U.S. Army's jet injector gun was a marked improvement over the single-point needle. Take rates were nearly 100 percent, and thousands of people could be brought together and vaccinated in one day. But although the foot-powered version of the jet injector did not require electricity, it was still a complex mechanical device that required frequent maintenance and was prone to breakdowns in the field. What was needed was a vaccination tool that was effective, yet simple enough to be used and maintained by thousands of unskilled health workers.

In 1961, Dr. Benjamin Rubin, a researcher with Wyeth Laboratories in Pennsylvania, and Gus Chakros, an engineer with the Reading Textile Machine Company, developed a bifurcated needle that could deliver small-pox vaccine into the skin cheaply and easily, with a minimum of training. The design of the needle was simple yet elegant: It consisted of a short length of steel wire hammered flat at one end and then punched to form two sharp-ened tines. When the needle was dipped into a vial of freeze-dried vaccine that had been rehydrated with a solution of 50 percent glycerine, a single drop of the viscous fluid was held by capillary action between the two tines of the needle.

Originally, Rubin intended the bifurcated needle for use in the same manner as the single-point needle. It was believed that unless the vaccina-tion was done gently, the resultant bleeding would wash out the vaccine virus. Henderson decided to see what would happen if the vaccinator held the needle perpendicular to the recipient's skin and made fifteen rapid punctures. He reasoned that because of the needle's construction, it would not penetrate deeply, and that it would be far easier to teach vaccinators to make forcible punctures rather than gentle "pressures." In fact, trials in Kenya and Egypt with vigorous vaccination, which caused a drop of blood to run down the recipient's arm, resulted in 100 percent satisfactory takes. Failures of vaccination were much more common when there was no evi-dence of blood at the vaccination site.

It was also found that the bifurcated needle gave superior results to the single-point needle with only one-quarter the amount of vaccine. Whereas each vial of freeze-dried vaccine had contained twenty-five doses when administered by the traditional method, the same vial could now provide up to one hundred doses when delivered with the bifurcated needle. As a result, the new vaccination method had the effect of greatly expanding the available vaccine supply. The small vials of freeze-dried vaccine and the packets of bifurcated needles were easily portable, making them ideal for field operations in which large numbers of vaccinators

traveled around rural areas by bicycle or on foot. In addition, the bifurcated needle was so easy to use that it was hard to make a mistake: Unskilled villagers could learn the technique by practicing for ten minutes
with an orange or a grapefruit. At collecting points where large groups of
people gathered, a single health worker could vaccinate about five hundred persons a day; going from house to house, it was possible to perform about 120 vaccinations a day.

Because of the clear advantages of the bifurcated needle, after 1969 it
replaced the heavier and more complex jet injector. Wyeth Laboratories
agreed to waive the licensing costs for the needles on the condition that the
WHO make them available only to developing countries participating in
the smallpox eradication campaign. To save money on mass production
of the bifurcated needle, the WHO shortened the shaft, reducing the cost
to about $5 per thousand. Manufactured from tempered steel, the needles
could be reused up to one hundred times without dulling. After a day of
vaccinating, field workers sterilized their dirty needles by placing them in
a pot of boiling water for half an hour. A superb example of "appropriate
technology," the bifurcated needle soon became the symbol of the WHO
smallpox eradication campaign.

Another important piece of the eradication puzzle was laboratory diagnosis. Most cases of smallpox could be diagnosed from clinical symptoms,
such as the pustular rash, with a high degree of accuracy. As the numbers
of cases in a country diminished, however, it was important to confirm the
diagnosis by distinguishing cases of smallpox from those of monkeypox
(which caused occasional human outbreaks in the rain forests of Central
and West Africa) or severe chickenpox. The main diagnostic method was
to extract the virus from a clinical specimen (such as a scab), cut a window
in the shell of a fertilized egg, and inoculate the virus onto the chorioallantoic membrane associated with the developing chick embryo. After a few
days of incubation, variola virus formed distinctive white, dome-shaped
pocks on the living membrane. In contrast, monkeypox virus produced

bright red pocks, caused by bleeding from tiny blood vessels, whereas chickenpox virus did not grow at all. An alternate diagnostic technique was to examine the viral isolate under an electron microscope. At high magnification, the biscuit-shaped particles of variola virus could easily be distinguished from the smaller, spherical particles of chickenpox virus.

Both of these diagnostic methods, however, were generally unavailable in the developing world. The egg membrane assay could not be performed in India, for example, because the local eggs were contaminated with other viruses that gave misleading results, and electron microscopes were too costly for poor countries to purchase and operate. At the same time, few virology laboratories in the industrialized countries were eager to take on the vital but unglamorous task of analyzing thousands of clinical samples from the smallpox eradication campaign.

Finally, two leading poxvirology labs, one at the U.S. Centers for Disease Control (CDC) in Atlanta and the other at the Research Institute for Viral Preparations in Moscow, agreed to divide up the diagnostic workload. Clinical specimens, such as scabs, were shipped on alternate weeks to Atlanta or Moscow for analysis. From 1967 to 1979, the two laboratories tested more than sixteen thousand specimens. Whenever a diagnosis of smallpox was made, a sample of the viral isolate was stored in a freezer for reference purposes. Over a period of several years, the CDC and the Moscow institute collected strains of variola major and variola minor from smallpox outbreaks in many parts of the world. It was therefore an historical accident that the two largest repositories of smallpox virus stocks came to be located on the territories of the opposing superpowers.

Another milestone on the road to smallpox eradication was the development of a new vaccination strategy. Before 1967, the prevailing approach was to vaccinate large masses of people indiscriminately. It was believed that if a large enough proportion of the population was rendered immune to smallpox, the resulting "herd immunity" would prevent the virus from spreading to new hosts and the disease would die out. Vaccinating 80 percent of the population over a five-year period was considered sufficient to interrupt smallpox transmission.

In densely populated countries such as India, however, smallpox persisted even in areas where, in strictly numerical terms, more than 80 percent of the population had been vaccinated. One problem was that mass vaccination campaigns often involved multiple revaccinations of easily accessible groups, such as schoolchildren, while failing to reach less accessible populations. Another problem was that in areas where the birth rate was high, the population received a constant influx of new infants, all of whom were susceptible to infection. In 1964, a WHO advisory committee concluded that 80 percent vaccination coverage was inadequate for smallpox eradication and recommended a goal of 100 percent, although 90 percent was probably sufficient. In densely populated countries, however, the large number of births every year made even the 90 percent target impractical.

The strategy devised by the WHO Smallpox Eradication Unit was to augment mass vaccination with a new approach known as "surveillance-containment." In the 1950s, Alexander Langmuir at the CDC had pioneered the concept of disease surveillance, meaning the early detection of outbreaks, but he had not included the active element of containment. Isolating smallpox patients and vaccinating everyone with whom they might come in contact served to create a ring of immunized persons around each outbreak, preventing the virus from infecting new hosts. In this way, it was possible to snuff out the flame of contagion before it could spread, much as building a firebreak could stop a forest fire in its tracks. The idea of sending out surveillance teams to find and contain outbreaks of smallpox as they occurred was thus an extension of Langmuir's idea.

A handbook published by the WHO Smallpox Eradication Unit in June 1967 called for a two-pronged strategy involving mass vaccination to cover 80 percent of the population, augmented with surveillance-containment programs in all countries with a relatively low rate of smallpox infection (fewer than five cases per 100,000 population). In countries with a higher rate of infection, surveillance-containment activities would be confined to major outbreaks and to areas where mass vaccination campaigns had already been carried out. The logic of combining the two approaches was that mass vaccination would lower the density of susceptible individuals in the population, reducing to a manageable level the number of smallpox outbreaks that the surveillance teams would have to trace and contain.

Soon after the AID/CDC vaccination program in West Africa got under way, the effectiveness of the surveillance-containment strategy was unexpectedly put to the test in Nigeria. Because of its large population and strategic location at the crossroads of region, Nigeria was key to the success of the campaign. Dr. William H. Foege directed the smallpox/measles vaccination effort in the southeastern part of the country, then known as Eastern Nigeria. A gentle giant at six feet, seven inches, Foege was a devout Christian with an abiding faith in the scientific method. From 1962 to 1964, he had trained in epidemiology, serving as an EIS officer in Colorado and obtaining a master's degree in tropical public health at Harvard University. In 1965 he had become a medical missionary with the Lutheran Church, which had assigned him to run a clinic in Ogoja province, one of three provinces in Eastern Nigeria. Foege and his wife lived in a village hut with dried-mud walls and no electricity or running water. They collected rainwater in a fifty-five-gallon drum and boiled it before drinking. Without even a fan to circulate the air, there was no escape from the debilitating heat and humidity.

After the White House approved the AID/CDC vaccination campaign for West Africa in November 1965, D. A. Henderson invited Foege to direct the effort in Eastern Nigeria with the permission of the Lutheran Synod. Although the program was not scheduled to start officially until January 1967, Henderson managed to get a few epidemiologists into the field by late 1966. Even so, shipments of four-wheel-drive vehicles and large amounts of smallpox vaccine would not arrive for a few more months.

On December 5, 1966, a local missionary called Foege by radio to report a suspected case of smallpox in his area. That afternoon, Foege visited the patient and confirmed the diagnosis. The question now was what to do about it. Smallpox vaccine was in extremely short supply—a few thousand doses, nowhere enough for a mass vaccination campaign—and most of it was in liquid form that required refrigeration. At 7:00 that evening, Foege got in touch by radio with the other Lutheran missionaries throughout the province. With a map spread out in front of him, he divided the area into zones and assigned one to each missionary. He then asked them to send runners to all the villages in their zones to determine whether cases of smallpox were present and to report back the following night.

By compiling the missionary reports, Foege identified several villages that had cases of smallpox. The extremely limited supply of vaccine forced him to rethink the WHO's vaccination strategy. Since the local language group numbered about 100,000 people and bordered on another major language group, vaccinating the entire population of Ogoja province was out of the question. Instead, Foege had the idea of focusing the initial campaign on just the smallpox-infected villages and the few areas where the virus was most likely to spread next. Because people living in villages without smallpox cases were not at immediate risk of infection, they could be vaccinated later on when fresh shipments of vaccine arrived.

That night, Foege and his two CDC colleagues sat around a kerosene lantern in his mud hut, studying the map of Ogoja province. They asked themselves: If they were variola viruses bent on immortality, where would they go next? Using information on market patterns, they identified three areas where the virus seemed most likely to spread. Accordingly, Foege sent out teams of health workers to begin vaccinating people in the infected villages and the three adjacent areas. The vaccinators were still hard at work on Christmas Day.

What happened next was truly remarkable. Only six weeks after the start of the vaccination campaign, no new cases of smallpox were reported in Ogoja province. Although less than 15 percent of the population had been vaccinated, transmission had been stopped dead in its tracks. It later turned out that cases of smallpox had been incubating in two of the three areas Foege had targeted. By the time the first rashes appeared, health workers had already saturated the area with vaccine, preventing any further spread. Foege and his colleagues had been fortunate to select the right places to vaccinate; otherwise they would have been behind the curve.

Thrilled by how quickly smallpox had been eliminated from Ogoja province, Foege reported his results to the Ministry of Health for Eastern Nigeria. He proposed dropping plans for a mass vaccination campaign and focusing strictly on surveillance-containment. Impressed by the success in Ogoja, the regional health authorities agreed to try out the new strategy.

In February 1967, working out of the regional capital of Enugu, Foege established a surveillance network throughout Eastern Nigeria that included both government and missionary sources. Whenever a smallpox outbreak was reported in a village, he dispatched a small team of health workers to isolate the sick individuals and to vaccinate everyone who was likely to come in contact with them. While searching for the source of an outbreak, the teams often identified other smallpox cases and vaccinated around them as well. During the first six months of 1967, health workers contained smallpox outbreaks throughout Eastern Nigeria by vaccinating some 750,000 people out of a total population of 12 million. Although this targeted strategy covered less than 7 percent of the population, it was sufficient to break the chains of smallpox transmission and rapidly reduce the incidence of the disease.

Meanwhile, ethnic tensions in Nigeria were approaching the boiling point. In January 1966, a bloody coup led by five army officers from the Christian Ibo tribe had toppled the civilian government in Lagos. The Muslim majority in Northern Nigeria, perceiving the coup as an Ibo conspiracy to seize control of the country, had responded with an insurrection in which they massacred tens of thousands of Ibos living in the north and burned their churches. Fleeing the violence, large numbers of Ibos migrated to their traditional homeland in Eastern Nigeria. On May 30, 1967, the region seceded from the rest of the country as the independent "Republic of Biafra," pushing Nigeria to the brink of civil war.

In early July, as health workers were containing the last known outbreaks of smallpox in Eastern Nigeria—now Biafra—the Nigerian government closed the border with the rest of the country. To attend a regional meeting of CDC advisers in nearby Ghana, Foege and his two colleagues had to cross the Niger River in a canoe. They were still in Ghana when hostilities broke out between Biafran and Nigerian government forces. Because of the fighting, Foege was unable to return to verify whether surveillance-containment had eliminated smallpox from Biafra.

After the civil war began, the vaccination campaign decreased in intensity but did not stop immediately. For several months, vaccine continued to be transferred from Nigerian to Biafran health authorities during cease-fire periods arranged for that purpose. The Nigerians would leave crates of smallpox and measles vaccine in the middle of a bridge and then withdraw and allow the Biafrans to pick them up. By the end of 1967, however, the vaccination program had ceased, and the following year a Nigerian government blockade caused a severe famine in Biafra, accompanied by epidemics of measles in refugee camps with high mortality. After the government of Biafra appealed for outside assistance in October 1968, the International Committee of the Red Cross and the World Council of Churches arranged to relaunch the vaccination campaign, which finally got under way in late December. During 1969, eleven teams administered 100,000 vaccinations a week, primarily against measles.

International relief agencies also conducted night flights into Biafra to deliver emergency food. On two occasions, relief workers found an individual with suspected smallpox and managed to send out scab specimens on one of the returning aircraft. Laboratory analysis revealed that both patients were infected with vaccinia virus, presumably as a consequence of vaccination. The fact that the relief workers had followed up two cases of suspected smallpox so carefully suggested that actual cases of the disease would have been detected. On that basis, Foege was reasonably confident that smallpox had been eliminated from Biafra.

The successful use of surveillance-containment in Biafra led the CDC to launch a major reform of the smallpox vaccination campaign throughout West and Central Africa in 1968. Known as "Eradication Escalation," the new strategy placed a greater emphasis on detecting and containing outbreaks. Some CDC epidemiologists were more enthusiastic about this approach than others, with medical officer Donald Hopkins perhaps the strongest advocate. Relying exclusively on surveillance-containment in Sierra Leone, the most heavily infected country in West Africa, Hopkins achieved "zero pox" in only nine months, from August 1968 to April 1969. Another CDC officer, Joel Breman, achieved similar results in Guinea.

Largely because of the epidemiology of smallpox in West and Central Africa, eliminating the disease from the region turned out to be far easier than had been anticipated. By reducing the number of susceptible individuals in the population through mass vaccination and then halting spread by finding and containing outbreaks, it was possible to interrupt smallpox transmission in one West African country after another with surprising speed. As of late 1969, the entire region had become free of smallpox—with the sole exception of Nigeria.

After nearly three years of fighting, the Nigerian civil war finally came to an end on January 15, 1970, when the Biafran forces surrendered in ignominious defeat. More than a million civilians had died, most of them from starvation and disease. After the cease-fire, Foege returned to Eastern Nigeria and determined that smallpox had indeed been eliminated from that area. Zero pox was achieved in the rest of Nigeria by May 1970. All the countries of West and Central Africa were now free of smallpox after only three and a half years of the planned five-year program. This impressive achievement helped to allay doubts about the feasibility of global eradication.

Meanwhile, surveillance-containment programs were also conducted effectively in other countries, with particularly successful results in two. In the Brazilian state of Paraná, which had a population of 8 million, WHO epidemiologist Ciro de Quadros, accompanied by a driver and two vaccinators, spent eight months tracing and containing outbreaks. After vaccinating fewer than fifty thousand people, the WHO team eliminated smallpox from the state. Equally dramatic results were achieved by a team of six vaccinators led by A. Ramachandra Rao in the south Indian state of Tamil Nadu, which had a population of 50 million people.

The WHO's initial two-pronged strategy of mass vaccination followed by surveillance-containment had been based on several assumptions that later proved to be incorrect. Although it had been believed that the endemic countries had a low level of herd immunity, the severity of smallpox meant that nearly all such countries had done some vaccination before the intensified WHO campaign began. As a result, it was usual to find that more than 50 percent of the population was already immune, either from vaccination or from natural disease. It had also been assumed that if a country had a high incidence of smallpox, the cases would be so widely scattered that a great many surveillance teams would be required to detect and contain the outbreaks. In fact, it turned out that most patients with smallpox transmitted the disease only to people with whom they came into close face-to-face contact, creating a chain of person-to-person transmission. Because smallpox patients tended to be clustered together rather than randomly dispersed, relatively few surveillance teams were needed to find each active case, trace the antecedent cases in the chain, and vaccinate all known or potential contacts.

Of the thirty-one countries where smallpox was endemic in 1967, eighteen had fewer than five cases per 100,000 population and thus should have begun incorporating the surveillance-containment method. Even so, many countries resisted the new strategy. Since mass vaccination was the traditional approach for controlling epidemic diseases, governments readily understood and accepted it. In contrast, the logic of finding and containing individual outbreaks was more difficult to grasp because it was labor-intensive yet involved vaccinating relatively few people. With vaccinators and vehicles always in short supply, it was hard to persuade national health officials that surveillance-containment operations should begin immediately and receive as high a priority as mass vaccination campaigns. Only gradually did the paradigm take hold that finding and containing smallpox outbreaks was more cost-effective. By 1970, most of the endemic countries had adopted surveillance-containment as an adjunct to mass vaccination, but it was only during the final phase of the WHO eradication campaign that it became the dominant strategy.

In April 1971, containment of the last smallpox cases in Brazil meant that the entire Western Hemisphere was now free of the disease. Henderson could hardly believe that smallpox was disappearing as fast as it was. In view of the rapid progress of the WHO eradication campaign, in October 1971 an advisory committee to the U.S. Public Health Service recommended halting the mandatory smallpox vaccination of American children prior to school entry because the risk of complications from vaccination now outweighed the risk of contracting the disease. Although no cases of smallpox had been reported in the United States in twenty-two years, six to eight deaths occurred annually from complications of vaccination. The rise in international air travel had slightly increased the probability of smallpox importations, but that danger was more than offset by the shrinking geographical areas where the disease remained endemic. Most smallpox victims lived in isolated rural areas or among the urban poor of developing countries and hence were unlikely to travel to the United States. Beginning in 1972, state public health departments gradually phased out the mandatory smallpox vaccination of children, although vaccination was still required for U.S. citizens traveling to countries where smallpox was endemic and for foreign nationals entering the United States. Other Americans who continued to be vaccinated were scientists working with vaccinia and other poxviruses, and military recruits.

*   *   *

Despite the important victories in the war against smallpox in West Africa and Brazil, the early 1970s saw a discouraging setback in the Middle East. Iran had eliminated smallpox in 1963 by means of a well-organized mass vaccination campaign. In October 1970, however, an Afghan family of six left their village near Kabul and went on a Muslim pilgrimage to the holy city of Mashhad in northeastern Iran. En route, they passed through a smallpox-endemic area of Afghanistan, where three of the children became infected. By the time the children were diagnosed with smallpox in early November, they had transmitted the disease to other pilgrims in Mashhad, triggering a major epidemic that spread throughout Iran over the next two years.

In January 1971, WHO headquarters in Geneva began to receive reports from travelers and foreign embassies about smallpox outbreaks in the Iranian capital of Tehran and in Tabriz, a major city. Alarmed, the Smallpox Eradication Unit sent a cable to Iran requesting information. About a month later, the Iranian government responded that nine cases of smallpox had occurred but that the outbreak was under control. Over the next several months, rumors of smallpox epidemics in Iran grew so numerous that the WHO made a second inquiry. This time, the Iranian government reported twenty more cases of smallpox—obviously a vast understatement. Frustrated, Henderson contacted some Iranian health officials he knew personally to obtain more information, but these individuals, fearful of the shah's secret police, declined to speak even confidentially about what they knew.

The reason for the cover-up was that the Iranian government was planning to hold a grand celebration on October 12, 1971, at the ancient city of Persepolis to commemorate the 2,500th anniversary of the Persian Empire. Heads of state and dignitaries from more than one hundred countries had been invited. Because smallpox was an embarrassment to the shah's regime and threatened to jeopardize the Persepolis event, Iranian officials had been ordered to suppress all news of the epidemic.

Meanwhile, the Iranian government launched a mass vaccination campaign, but because the country's indigenous stocks of liquid vaccine were of poor quality, fewer than 65 percent of the recipients were successfully immunized. Thus, although nearly 20 million people were vaccinated in 1971, the disease continued to spread. Major outbreaks involving thousands of cases occurred in the city of Shiraz in southwestern Iran, only thirty miles from Persepolis, where the celebration was to take place. The Iranian health authorities found themselves in a terrible dilemma: Despite their desperate need for more vaccine, they could not issue a formal request to the WHO because of the cover-up that had been ordered at the highest levels.

Finally, in November 1971, the Iranian deputy minister of health contacted the WHO Smallpox Eradication Unit and pleaded for assistance. He had gone to Shiraz to oversee the vaccination of schoolchildren and had personally administered the vaccine to a little girl. Because the vaccine was defective, the child had subsequently contracted smallpox and died. Stricken with guilt over this tragedy, the deputy minister requested an emergency shipment of freeze-dried vaccine, which Henderson agreed to provide. The WHO shipped 2 million doses to Iran in late 1971, followed by 11 million more in 1972. This time, the vaccination campaign was effective and the last case of smallpox in Iran was reported in September 1972.

By then, however, the epidemic had spread to other countries of the region. In late 1971, travelers infected with smallpox had crossed the border into Iraq, carrying the disease to Baghdad and then along the main highways to Arbil in the north and As Samawah in the south. Although the outbreak in Iraq involved at least eight hundred cases, the government reported fewer than forty. In February 1972, smallpox spread to Syria. The government in Damascus reported a total of fifty-four cases and two deaths, far fewer than the actual number, and managed to contain the outbreak by June 1972.

In August, after the epidemic in Iran had subsided, the shah's government finally allowed D. A. Henderson to visit the country. He was accompanied by Ahmad Hajian, an Iranian native based at the WHO's regional office in Alexandria, Egypt, who served as his interpreter. The two men traveled through areas of northwestern Iran where serious outbreaks of smallpox had occurred. At each hospital and clinic they visited, it was obvious that the doctors and administrators had rehearsed their answers in advance. When Henderson asked how many cases of smallpox had been seen recently, his Iranian interlocutor would glance furtively around the room to see who else was listening. The usual response was that there may have been a few cases, but he really didn't know how many.

During a visit to the central hospital in Tehran, Henderson toured an enormous smallpox ward with about one hundred empty beds, although he had been told repeatedly that there had been only thirty-one cases of smallpox in the entire country. "Why do you have all these beds?" he asked.

"We want to be prepared in case a big outbreak occurs," the hospital director explained.

Henderson feigned surprise that Iran would spend such a large amount of money to prepare for a disease that was apparently of such minor importance. He then asked how many patients had died in the recent epidemic. None, the Iranian doctor replied. Disgusted at this obvious lie, Henderson commented sarcastically that the Iranian government should publish its remarkable results in *The Lancet,* a leading British medical journal. At this point one of the government escorts became agitated and dropped back to converse with Hajian in Farsi. "He's making fools of us," he whispered. "Of course we had more cases and deaths, but we can't admit it."

Based on his travels in Iran, Henderson estimated that the epidemic had involved roughly eight thousand cases of smallpox, of which about two thousand had been hospitalized. Several years later, in December 1978, the Iranian government submitted a confidential report to the WHO providing information on 1,996 cases in 400 scattered outbreaks between November 1970 and September 1972. But Henderson remained convinced that Iran had greatly understated the actual number of cases.

*   *   *

In early 1972, the smallpox epidemic in the Middle East spread unexpectedly to Europe. On February 15, Ibrahim Hoti, a thirty-eight-year-old ethnic
Albanian Muslim from the Yugoslav province of Kosovo, returned to his
home village after a six-week pilgrimage, or hajj, to the holy shrine of Mecca
in Saudi Arabia. A clergyman in a semisecret Muslim sect, he had traveled in
a chartered bus with twenty-five other believers from Kosovo. On their way
home, the pilgrims had toured Muslim holy sites in southern Iraq and then
spent three nights in Baghdad, where smallpox was present.

Although smallpox vaccination was still compulsory in Yugoslavia, the
country had not experienced a case of the disease in more than forty years.
As a result, the public health authorities had become complacent and immunization requirements had been relaxed. Vaccination coverage was particularly spotty in the autonomous southern province of Kosovo, which had
poor roads and public services.

The morning after he returned to his home village, Hoti fell ill with
symptoms consisting of fever, fatigue, shivers, and body aches, which he
attributed to the long bus journey. A few days later, he developed a sparse
red rash, but he recovered quickly and did not seek medical help. None of
his fellow travelers became sick. Hoti had been vaccinated against smallpox two months earlier and may have had partial immunity. In any event,
he acquired an extremely mild case; a month later, no trace of the rash could
be seen on his face or body.

For two weeks after Hoti's return from the pilgrimage, his relatives came
to visit from all over Kosovo to hear about his travels. He also met with
friends at a tavern in the nearby town of Djakovica. A few weeks later, eleven
people traceable to Hoti came down with fever and rash. Six of the victims
were from his village, one was a relative from another town who had visited him at home, and four were strangers whom he had apparently infected
at the market in Djakovica.

One of the four strangers was Ljatif Muzza, a thirty-year-old teacher
who had come to Djakovica on February 21 to enroll at the Institute of

Higher Education. On March 3, he developed a fever, and two days later he went to a local clinic, where doctors suspected a bacterial infection and gave him penicillin. The drug had no effect, however, and over the next few days, Muzza became critically ill. On March 7, after dark bruises had appeared on his skin, his brother took him by bus to a district hospital about one hundred miles away. There, Muzza's condition continued to deteriorate, and two days later he was transferred by ambulance to the main hospital in Belgrade. By then, the bruises on his skin had turned almost black. During hospital rounds, the attending physician presented him to medical students and staff as a case of severe allergic reaction to penicillin.

On March 10, Muzza began to bleed uncontrollably into his intestines, expelling quarts of blood that stained the bedsheets black. He was rushed to an intensive care unit, where he died twenty-four hours later. The duty physician listed the cause of death as an allergic reaction to penicillin. No one suspected that Muzza had hemorrhagic smallpox, or "black pox." In addition to being invariably lethal, this rare form of the disease caused patients to emit large quantities of aerosolized variola major virus, creating an extreme risk of contagion. During his weeklong odyssey, Muzza had seeded smallpox across Yugoslavia. By the time of his death, he had infected thirty-eight other people, most of them health care providers and other patients in the hospitals he had visited.

By mid-March, Yugoslavia was experiencing a full-blown smallpox epidemic. The first wave of 140 cases had been reported in twenty-five villages and towns in Kosovo province. On March 16, Yugoslavia's authoritarian Communist government, headed by Josip Broz Tito, declared a state of emergency and launched draconian containment measures. Army troops built roadblocks to seal off the infected villages and towns and banned all public meetings. They seized numerous hotels and residential apartment blocks, ringed them with barbed wire and armed guards, and quarantined about ten thousand suspected contacts for two weeks or more. The quarantined individuals had their temperatures taken and their skin inspected for rashes on a daily basis.

Meanwhile, more than three hundred health teams, some of them army medical units, began mass vaccination campaigns in concentric rings around the twenty-five infected villages and towns, so that by the end of the month almost the entire population of Kosovo province had been vaccinated. Because of a shortage of needles, many people were vaccinated with pens or styluses. Use of liquid vaccine also resulted in unsuccessful takes, so that a large proportion of the vaccinations had to be repeated. Meanwhile, neighboring countries closed their borders with Yugoslavia and advised their citizens not to visit, and shipments ranging from vegetables to cement were stopped at the border.

On March 20, Muzza's brother developed a smallpox rash. Only then did the Yugoslav medical authorities recognize in retrospect that Muzza had died of hemorrhagic smallpox. Because of the difficulty of tracing his thousands of possible contacts, the Yugoslav authorities decided on March 24 to vaccinate the entire population. In response to an urgent appeal, thirteen countries contributed more than 15 million doses of vaccine. Over a period of four days, 1.2 million people in the Belgrade area were vaccinated at some 2,500 vaccination posts manned twenty-four hours a day. Within three weeks, 18 million people throughout Yugoslavia were vaccinated, out of a total population of 20.8 million.

As the countrywide vaccination campaign got under way, a third wave of twenty-three smallpox cases appeared in Kosovo and Belgrade between March 31 and April 12. Fears that the disease might spill over to other European countries grew when a Yugoslav migrant worker was diagnosed with smallpox in Hannover, West Germany. In the United States, public health officials placed more than one thousand visitors from Yugoslavia under medical observation. Although seventeen of them developed rashes or other signs of illness requiring further study, no cases of smallpox were diagnosed.

Thanks to the vaccination of nearly the entire population of Yugoslavia, and the fact that roadblocks and quarantines prevented infected people from traveling, a fourth wave of cases did not fully materialize. The last victim, a ten-year-old boy, was hospitalized on April 12, two months after Ibrahim Hoti's return from Iraq. In all, the epidemic had involved 175 cases of smallpox and thirty-four deaths.

As a harrowing example of the dangers posed by a single imported case of smallpox, the Yugoslav outbreak demonstrated that no country would be safe until the disease had been eliminated from all corners of the globe. Achieving that objective was by no means certain, however, for the greatest challenges of the smallpox eradication campaign still lay ahead.

LONG ROAD TO ZERO

The next phase of the smallpox eradication campaign focused on the Indian subcontinent. Because of its vast population and high rates of infection, this region posed enormous challenges: India alone accounted for nearly 60 percent of the world's reported cases of smallpox. The disease also behaved differently in South Asia than in other endemic areas. In West Africa, many people were unvaccinated, but the disease spread slowly because rural areas were sparsely settled and villagers did not travel over long distances. In India, by contrast, more than three-quarters of the population had some immunity to smallpox because of past disease or vaccination, but the sheer numbers of people created a density of susceptible hosts higher than anywhere else in the world. The mobility of the population was also enormous: At any given moment, as many as 6 million Indians were moving from one place to another. Most of them traveled on 10,800 daily trains—with the poorest riding without tickets on the tops of rail cars—or on public buses that plied the extensive road network.

Another characteristic of South Asia was that the strains of variola major that circulated there in the late 1960s and early 1970s were more than twice as deadly as those found in West Africa. A possible explanation is that in rural Africa, where people lived in scattered, low-density settlements, evolutionary pressures had caused the smallpox virus to become milder so that infected individuals would survive long enough to transmit the virus to another susceptible host. In the crowded cities of South Asia, in contrast, individuals infected with smallpox readily spread the disease to others. Because the virus had numerous opportunities for transmission, more lethal strains did not pay a high evolutionary price for killing their hosts.

Beyond the practical and logistical hurdles to smallpox eradication in South Asia, the WHO faced major cultural and psychological obstacles. Smallpox had existed on the subcontinent for well over three thousand years and had become deeply woven into the patterns of daily life. In India, the disease was so pervasive that it had its own deity, known in the north of the country as Shitala Mata, or "cooling mother." Worship of the smallpox goddess dated back at least to the twelfth century. People made offerings at shrines dedicated to Shitala Mata and to images in the home, and annual festivals were held on her feast day in March, near the peak of the hot, dry smallpox season. Although specific beliefs about the smallpox goddess varied from place to place, she was usually endowed with the power both to inflict the disease and to prevent or cure it.

In Hindu tradition, Shitala Mata was the wife of Shiva the Destroyer. She was portrayed in paintings and sculptures as a large-eyed goddess who traveled about on the back of a lactating female donkey, carrying a broom and a water pot and sowing poisonous grains that caused the pustules of smallpox. A patient's fate depended on whether she used the cooling water or the ineffective dry broom to remove the poisonous grains. Hindu shrines to Shitala Mata usually consisted of a small temple with a bell on top and an image of the goddess riding on a donkey. In front of the shrine was a neem tree, a symbol of the disease because it bloomed during the peak smallpox season. Donkey milk, supposedly a cure for smallpox, was often sold outside.

According to belief, when Shitala Mata was flattered with pleasant offerings and cooled with fragrant baths, she withheld her rage and spared the worshiper, but when angered, she inflicted the burning fever and rash of smallpox. Village women attempted to lure her away from their children by enticing her with pots of water and plates of cooling foods, such as cold rice, plantains, and yogurt, which they placed on rooftops for her to enjoy. Some believers resisted vaccination because they feared that defying the will of the goddess would provoke her wrath.

The cult of Shitala Mata reflected a pervasive fatalism about smallpox as an unchangeable fact of Indian life. Middle-class Indians viewed the disease as a bane of the impoverished lower castes, who could not afford to be immunized and lacked access to the most basic health services. Even some Western epidemiologists believed that South Asia was the cradle of smallpox and that the disease would never be eliminated from the region.

At least initially, these pessimistic assumptions were called into question by the surprising success of efforts to eliminate smallpox from East Pakistan. The government there conducted mass vaccination campaigns in 1961–63 and 1968, followed by the introduction of a surveillance-containment program in 1969. In August 1970, local surveillance teams and WHO advisors scoured the countryside looking for smallpox cases, and found none. The sudden disappearance of the disease from East Pakistan was wholly unexpected. For Henderson and other WHO officials, the news was almost too good to be true.

Several months later, however, a tragic setback occurred. Tensions had been building for years between East and West Pakistan, which were separated by one thousand miles of Indian territory and vast economic and cultural disparities. Although more than half the country's population lived in the smaller eastern enclave, West Pakistan controlled most of the economic and political power. When East Pakistan's demands for greater autonomy were rejected, it seceded from the union and declared itself the independent state of Bangladesh on March 16, 1971. In response, West Pakistan mobilized its army and civil war broke out.

During the fighting, some 10 million East Pakistanis fled across the border into the Indian state of West Bengal, where about 250,000 of the refugees were housed in the teeming Salt Lake Camp outside Calcutta. The state health authorities were too busy meeting the basic needs of the refugees to conduct systematic vaccination campaigns, and they denied repeated requests by WHO representatives to visit the camp. Because smallpox was still endemic in West Bengal state, the disease was apparently introduced into the refugee camp in November 1971. Although several refugees developed facial rashes, these cases were misdiagnosed as chickenpox.

Thanks to the intervention of Indian troops during the final weeks of the civil war, Bangladesh won its independence on December 16, 1971. Within a month, an estimated fifty thousand refugees returned home, and Salt Lake Camp prepared for a major exodus. On January 19, 1972, an epidemiologist at the CDC in Atlanta saw a video clip of the camp on the television evening news and recognized smallpox rashes in close-ups of some of the refugees. Alarmed, he telexed Henderson in Geneva, who in turn notified the WHO field office in New Delhi that a smallpox epidemic might have broken out. The New Delhi office then telexed the West Bengal government in Calcutta. Despite denials from state officials that anything was wrong, the WHO obtained permission for Dr. S. N. Ray of the Indian national program office to travel to Salt Lake Camp and investigate the situation. Dr. Ray flew to Calcutta and arrived at the camp one day before most of the refugees were to be sent home. Discovering, to his horror, that smallpox was rampant, he organized an emergency vaccination campaign, but by then it was already too late. Because the camp was loosely organized, a tidal wave of refugees began departing over the next several days. Individuals with active smallpox, others incubating the disease, and unvaccinated people were loaded together on trucks, trains, and buses for the trip to the border.

Tragically, the flood of returning refugees reintroduced smallpox to the new state of Bangladesh. In March 1972, the disease reached Khulna City and spread like wildfire through the densely populated slums, or *bastis*. Over the next few months, the WHO country team tried desperately to contain the epidemic by hiring some four thousand temporary vaccinators. In July and August, however, a drought in the countryside destroyed most of the hastily planted crops, forcing hungry villagers to migrate to the slums of Dacca, the Bangladeshi capital, and to several shelter camps, widely dispersing smallpox in the process. It took the WHO eradication campaign nearly four years to recover from this devastating setback.

By the end of 1972, twenty-five countries where smallpox had been endemic in 1967, including Indonesia, had reached the goal of "zero pox," but several others still faced formidable challenges, particularly India, Bangladesh, Pakistan, and Ethiopia. In India, the WHO country office was located in the capital, New Delhi, in a modern five-story office building on the bank of the Jumna River near the Red Fort. Directly across the river, a coal-fired power plant belched thick black smoke in ironic counterpoint to the name and mission of the World Health Organization.

Heading the four-person WHO country team was Dr. Nicole Grasset, a Swiss-French virologist and epidemiologist in her forties. The daughter of a famous Swiss microbiologist, she had grown up in South Africa and, after studying medicine, had worked at the Pasteur Institute in Paris. In 1969, during the civil war in Nigeria, she had hand-carried measles vaccine to Biafran villages for the International Committee of the Red Cross. The attractive, dark-haired Grasset had a French sense of style and was always elegantly dressed, even when she ventured into the bush. She was warm, sophisticated, and tough-minded, and her passionate dedication to the cause of smallpox eradication drove her to work eighteen-hour days.

The youngest member of the country team was Dr. Lawrence Brilliant, a twenty-nine-year-old American physician. In 1972, he and his wife Girija had moved from California to India to live in a Hindu monastery in the foothills of the Himalayas, planning to devote their lives to spiritual study and meditation. Brilliant grew a bushy black beard and long hair that hung down to the middle of his back, and he wore a *kurta dhoti*, or pajama, made of Indian homespun cloth. One day his guru, Neem Karoli Baba, told him to return to the world and use his medical training to help eliminate smallpox from India. Brilliant made the ten-hour journey by bus, train, and rickshaw to the WHO office in New Delhi to offer his services, but Nicole Grasset turned him down because of his youth, inexperience, and unconventional appearance. She wrote him off as just another hippie and, in any case, no medical jobs were available. At the insistence of his guru, Brilliant returned several more times to the WHO office, each time dressing a bit more conservatively. Impressed by his persistence, Grasset finally hired him as an administrative assistant because he could type and spoke fluent Hindi.

When Brilliant arrived in New Delhi with his wife in June 1973, he carried all their possessions in a small suitcase. Despite his modest title, he soon took on greater responsibilities with the smallpox program. He made good use of his fluent Hindi by meeting with Indian spiritual leaders and persuading them to support vaccination, and he also had the clever idea of setting up surveillance posts at shrines to Shitala Mata, where people suffering from smallpox often went to pray.

The other two members of the WHO country team in India were Zdeno Jezek and William H. Foege. Jezek, a dynamic Czech epidemiologist, had come to steamy India after working for five years on the chilly steppes of Mongolia, where he had set up that country's first disease control system. Foege, the American physician who had given the surveillance-containment strategy its first real test in Biafra, had returned to the CDC in Atlanta when the West African campaign was over. On learning of the daunting challenges facing the smallpox program in India, however, he decided to go back to the field. Having arranged for the CDC to second him to the WHO, he arrived in New Delhi in August 1973. In addition to Foege's deep understanding of epidemiology, he had a wry sense of humor. One day he visited an Indian village where a surveillance team was having trouble persuading the local children to come out and be checked for smallpox vaccination scars. A towering presence at six feet, seven inches, Foege asked the village chief to announce that "the tallest man in the world" was visiting. Soon he was surrounded by a crowd of curious children.

Five Indian government officials also played a key role in the smallpox campaign. In addition to Dr. P. Diesh, the assistant secretary for health, and Dr. M. I. D. Sharma, a warm, grandfatherly gentleman who headed the National Institute of Communicable Diseases, three Indian health experts worked in the WHO country office in New Delhi. Grasset, Foege, Jezek, and Brilliant developed a strong bond of trust with their Indian counterparts, who were all well educated and hard-working, and they often sat down together and made decisions collectively.

The Indian government had launched its own campaign in 1962 to eliminate smallpox from the country, hiring large numbers of vaccinators and building several vaccine production facilities with the intention of stopping transmission within five years. In 1967, however, India recorded its largest number of smallpox cases in a decade. Discouraged and frustrated, Indian health officials came close to abandoning the idea of a special smallpox program. In addition to serious management problems, the Indian campaign had been handicapped by the use of liquid vaccine or substandard freeze-dried vaccine that deteriorated in the heat, and a problematic delivery system. Health workers administered the vaccine with a rotary lancet, a stylus with a small flat plate mounted on the end from which five sharp prongs extended. Drops of vaccine were placed at four sites on the recipient's skin and the lancet applied and vigorously twisted, an extremely painful process that caused many people to fear and avoid vaccination. Moreover, the wound caused by the rotary lancet often became infected and left a permanent scar, making it impossible to determine whether or not the vaccination had taken.

The WHO campaign, in contrast, employed good-quality freeze-dried vaccine and delivered it with the bifurcated needle, which was much less painful and traumatic. Nicole Grasset also recognized that although mass vaccination could reduce the incidence of smallpox, the active detection and containment of outbreaks was essential to eliminate the remaining foci of infection. Even if it were possible to vaccinate all of India's 600 million people, the vaccinators would be unable to cope with each year's new cohort of 25 million babies, all of whom were susceptible to infection.

Although the effectiveness of the surveillance-containment strategy had been demonstrated in West Africa and elsewhere, it had to be adapted to the special conditions prevailing in India. One problem was that the reporting of smallpox cases by local hospitals and clinics was highly unreliable. Because senior government authorities considered the continued presence of smallpox in an area as evidence of poor performance, district and state health officials tried to save face by reducing the number of cases they reported up the chain. As a result, less than one-tenth of all smallpox cases in India were made known to the Ministry of Health. Another difficulty arose from the high mobility of the population: It was not uncommon for an entire family of landless peasants to travel to an urban area in search of seasonal work and then return to their home village when a family member developed smallpox. Because of such frequent migrations, surveillance teams could not move fast enough to detect outbreaks and prevent them from spreading.

Without reliable data on the actual incidence of smallpox and the locations where most of the cases were occurring, the members of the WHO country team in New Delhi felt as if they were groping in the dark. To overcome the poor voluntary reporting of smallpox cases, in the summer of 1973 they developed a new strategy, which they called "active search." Local health workers would visit all the towns and villages in a geographical area to ferret out the hidden cases of smallpox, after which separate containment teams would be dispatched to vaccinate around each outbreak and break the chains of transmission. Because large numbers of government health workers were employed at health centers and district offices throughout India, enough manpower was available to search extensive areas of the countryside in a short amount of time.

Grasset decided to schedule the active searches for the fall because it was the annual low point of smallpox incidence in India. The number of cases fluctuated yearly according to seasonal patterns, with the period of maximum transmission during the cool, dry winter months from December through May (usually reaching a peak in April), and the trough during the hot, humid monsoon season from June through November. The main reason for this seasonal fluctuation was the effect of temperature, humidity, and exposure to ultraviolet radiation on the survival time of variola virus in the air. When the weather was cool and dry, the airborne virus could survive for as long as twenty-four hours, greatly increasing the potential for person-to-person transmission, but when the weather was hot and humid, the virus died off in a matter of minutes. Another reason for the higher incidence of smallpox during the winter months was that it was the peak season for weddings and fairs, which were often attended by large crowds.

The WHO team hoped that by searching for smallpox cases in the fall, during the annual low point of transmission, it would be possible to break the back of the disease before the winter months arrived and the number of cases began rising steeply again. Nicole Grasset planned to focus the initial searches in northern India. By 1973, 93 percent of all cases of smallpox in the country were concentrated in the four large northern states of Uttar Pradesh, Madhya Pradesh, Bihar, and West Bengal. Together, these four states had 249 million inhabitants, or slightly more than the entire population of the United States at the time.

The WHO office in New Delhi spent the months of July and August 1973 preparing for the active search operations in the fall. Working in the oppressive 120-degree heat and high humidity of an Indian summer, the international advisors and their local colleagues developed new procedures, trained health workers, designed reporting forms, and produced posters and "recognition cards" with color photos comparing the rashes of smallpox and chickenpox at various stages.

During three "search weeks" in October, November, and December 1973, an army of health workers fanned out across the four northern states to search for smallpox cases in thousands of villages and urban districts. Twenty-two surveillance teams were deployed, each responsible for covering an area inhabited by some 10 million people. When the search of a village turned up an active case of smallpox, a containment team was sent in to isolate the patient and vaccinate the family members and the twenty to thirty adjacent households.

The active search campaign revealed that the actual incidence of smallpox in northern India was much higher than had been assumed. More than thirty thousand cases were discovered between October and December 1973, almost five times as many as had been reported voluntarily during the same period in 1972. In Bihar state alone, more than one thousand villages were infected with smallpox. The WHO was stunned by the magnitude of the epidemic, and containment teams were unable to keep up with the large number of outbreaks being reported.

Out in the field, the dogged work of searching for smallpox cases and vaccinating contacts went on, day after day. The dedication of the international advisers was extraordinary. They lived in small rural towns without electricity or fans, ate the local cuisine, and worked eighteen-hour days, seven days a week. Typically they each lost between ten and twenty pounds during a three-month tour of duty. At the same time, though, the psychological rewards were great. The foreign advisers and their local counterparts were driven by a shared sense of idealism that transcended differences of culture, race, religion, and nationality. They had caught "eradication fever"—the dream that a horrible disease could be vanquished forever through their collective efforts—and they were prepared to sacrifice family life, career ambitions, and even personal health in a single-minded commitment to that goal.

Once a month, the international advisers met with Indian government officials in the state capitals to report on progress and discuss ways of improving their tactics. The hotels where the meetings took place were usually air-conditioned, the only relief from the heat and humidity the advisers would have all month, but they were so absorbed in their work that they disliked being away from the field for even a few days. Although frequently on the verge of physical and mental exhaustion, they were driven forward by a deep sense of urgency: Areas recently freed of smallpox could easily be reinfected, causing their hard-won victories to unravel. And if they failed, the world might never again be prepared to join across political boundaries to fight a common scourge.

In January 1974, Henderson met in New Delhi with the four senior members of the WHO country team, all of whom had lost weight and looked exhausted. Working under primitive conditions in the field, they had contracted various ailments. One had an incapacitating kidney infection, the second a painful case of shingles, the third a serious fungal infection in one foot, and the fourth a case of pneumonia with high fever. Yet despite their obvious discomfort, the sole topic of discussion was how to find the additional resources needed to sustain the momentum of the smallpox campaign. When Henderson finally asked how the team members could keep working under the circumstances, one of them replied that things could not get any worse and would therefore have to get better.

Late March, however, brought more discouraging news. Reports of smallpox cases began arriving from parts of Madhya Pradesh state that had earlier been freed of the disease. Investigation revealed that the outbreaks were the result of importations from the nearby state of Bihar. The source of infection was traced to Jamshedpur, an industrial city with a population of about 800,000. Commonly known as Tatanagar, it was the seat of Tata Industries, India's largest private corporation. With its vast complex of iron and steel mills, Tatanagar was the Pittsburgh of India. The city was linked by extensive road and rail networks with the rest of the country, and tens of thousands of landless workers traveled there in search of seasonal employment.

In late April 1974, Nicole Grasset sent Larry Brilliant to assess the smallpox epidemic in Tatanagar, an assignment that made him both proud and anxious. He had previously worked in Madhya Pradesh, which had a relatively low population density and well-developed health services. But Bihar was a different story: thickly populated, lawless, chaotic, and the most impoverished state in India. Until then, Brilliant had been the junior member of the team, but now he had to dig deep within himself to find the courage to act.

When he arrived in Tatanagar, the scale of the smallpox epidemic dwarfed any other since the start of the Indian campaign. More than a hundred smallpox patients had flooded the city's fifty-bed hospital, and the local high school had been converted into an emergency quarantine center. At the maternity hospital, dozens of pregnant women had bled to death from hemorrhagic smallpox. As Brilliant walked around the city, he heard graphic stories about the horrors of the epidemic. A month earlier, a raft of dead bodies had choked the river flowing through Tatanagar, and vultures had flown overhead carrying tiny limbs spotted with pustules. At a Tata housing complex called the Kaiser Colony, Brilliant saw the bodies of dead children lying in the gutters and grieving mothers begging for assistance. The scene was etched in his brain like Hieronymous Bosch's vision of hell. Smallpox was such a repulsive, awful disease that it made Brilliant angry at God.

It soon became apparent that Tatanagar was exporting smallpox to parts of India and Nepal that had only recently been freed of the disease. Temporary mill workers were contracting the disease and returning by train to their native villages, where they spread the infection further. More than two thousand cases and five hundred deaths in eleven Indian states had been linked to travelers from the Tatanagar railway station. Western pilgrims from London and Tokyo had also been infected with smallpox during visits to Bodhgaya, a nearby Buddhist shrine.

As the epidemic in Bihar continued to intensify, the mood of the WHO country team turned grim. The weather was unbearably hot, often reaching 120 degrees with 80 percent humidity; the peak smallpox season had begun, and the number of outbreaks was climbing relentlessly. Just when thousands of health workers were needed to search for smallpox cases and to contain outbreaks, half the vaccinators in Bihar went on strike and the government physicians threatened to join them. Air India and railroad workers staged work stoppages, nearly halting the shipments of smallpox vaccine from New Delhi to Bihar, and gasoline was in short supply.

Meanwhile, India was preparing to explode its first atomic device—officially, for "peaceful" purposes—at a test site in the Rajasthan desert. Journalists from all over the world converged on New Delhi to report the event. On May 18, the day of the nuclear test, the smallpox epidemic in Bihar achieved an appalling record, with 8,664 infected villages and 11,000 individual cases reported in a single week. The foreign press could not resist the irony of a country with the technological prowess to explode a nuclear device simultaneously at the mercy of a Stone Age disease. A large crowd of reporters gathered outside the WHO office building in New Delhi, but no senior staff members were available for comment: Grasset was hospitalized with a kidney stone, Foege was in Bihar, and Jezek was in southern India, leaving Brilliant nominally in charge. He attempted to answer the reporters' questions but, feeling out of his depth, he called D. A. Henderson in Geneva and asked for assistance. Henderson and Arita flew immediately to New Delhi and showed Brilliant how to work with the press.

Also caught in the glare of the media spotlight were Indian Minister of Health Karan Singh and Prime Minister Gandhi. When international attention began to focus on the smallpox epidemic at the expense of the successful nuclear test, members of the Indian Parliament angrily demanded an explanation. For years, the Indian government had resisted Henderson's requests to expand the WHO program by bringing in more foreign epidemiologists. After the nuclear test, however, the New Delhi office was authorized to hire four more international advisers.

For the rest of May and into June, the epidemic in Bihar continued to spread, with more than eight thousand new cases reported each week. Morale was deteriorating, and Indian government officials began to express serious doubts about the effectiveness of the surveillance-containment strategy. During a meeting with the WHO country team, Bihar's minister of health advocated a return to mass vaccination. Senior WHO staff members protested in vain until a young Indian physician working in one of the districts pointed out that when a house was on fire, it was the usual practice to pour water on the burning house and not on all the houses in the village. Swayed by this argument, the health minister agreed to delay a decision for a month. If no apparent progress had been made by then, he would insist on a return to mass vaccination.

Larry Brilliant was increasingly convinced that something drastic had to be done to contain the epidemic in Tatanagar. Although the city administration was in the hands of Tata Industries, company management was doing nothing to address the crisis. Late one night in early June 1974, Brilliant decided impulsively to visit Russi Mody, the managing director of the Tata Iron and Steel Corporation. Around midnight, he drove his jeep to Mody's house in a wealthy district of the city. A large German shepherd guard dog began barking fiercely, and a servant came out and escorted Brilliant inside. When he entered, Mody was having a late dinner with his deputy, Sujit Gupta. Not even pausing to introduce himself, the bearded young American blurted out, "Do you have any idea what you're doing to the world?"

Mody was taken aback, but he agreed to listen to what Brilliant had to say. The three men talked until dawn. In the course of their conversation, Mody went from being unaware of the smallpox epidemic in Tatanagar to becoming excited about the WHO eradication campaign. He assigned Gupta, his number-two man and a dynamic young Bengali, to work with Brilliant. The next day, Mody sent a telex to the company president, J. R. D. Tata, informing him about the situation.

A few weeks later, Brilliant and Grasset met with Tata at his home in New Delhi. At first, the wealthy industrialist was skeptical. For him, smallpox in India was more of a state of nature than a disease, and the concept of eradication was completely foreign to him. But when Grasset explained the scientific basis of the WHO campaign with great passion and conviction, Tata was won over. He and Grasset negotiated an agreement under which the WHO would provide technical assistance and Tata Industries would donate $500,000 worth of materials, manpower, and vehicles to help fight smallpox in southern Bihar. The company mobilized fifty doctors, two hundred paramedical supervisors, and more than nine hundred workers to search Tatanagar and the 1,760 villages within a forty-five-kilometer radius to find and contain outbreaks.

To prevent the epidemic from spreading further, Brilliant placed all of Tatanagar under quarantine. As a United Nations employee, he had no legal authority to close the city, but he did so by force of will, inspired by idealism and youthful daring. He ordered teams of vaccinators to erect barricades at all the bridges and major roads leading into the epidemic zone, refusing anyone permission to enter or leave without proof of vaccination. Checkpoints were established at bus stations, and trains were rerouted away from the central train station—a major focus of the epidemic—and diverted to special platforms where passengers could be examined for vaccination scars as they got on and off the cars.

Tata Industries and the WHO epidemiologists worked systematically to contain the smallpox epidemic in Tatanagar and southern Bihar.

A local professor of demography named Dr. Vishwakarma had done his Ph.D. dissertation on the effects of urbanization on rural development and had made detailed maps of all the villages in the region. The WHO advisers used these maps to determine where smallpox was most likely to spread and to plan their containment strategy. Vishwakarma was delighted that his academic research was being put to some practical use.

One by one, the outbreaks were detected and snuffed out. Bihar had a well-earned reputation as a particularly lawless state, and one of the villages infected by smallpox was inhabited by numerous *dacoits,* or armed thieves, who pillaged the countryside. A health worker who entered this village to vaccinate was robbed of his possessions and stripped naked. That night, several Indian policemen armed with shotguns accompanied a team of vaccinators and guarded them while they worked.

Within a few months, the WHO and Tata Industries had made impressive strides in containing the outbreaks in southern Bihar. One of the last major hurdles was a village on the outskirts of Tatanagar that was infected with smallpox but was resisting vaccination for religious reasons. The village was headed by Mohan Singh, a forty-year-old spiritual leader of the Ho tribe who preached that it was contrary to God's will for humans to attempt to prevent or cure any disease they might contract. Health workers in the surrounding district had been making good progress in containing smallpox until Mohan Singh's village began spreading the disease into neighboring areas.

On several occasions, Larry Brilliant went to the village and attempted to persuade Mohan Singh to accept vaccination. As a spiritual man himself, he explained that it was his *dharma,* or religious duty, to protect the village children from contracting smallpox. If even a single focus of the disease was left to smolder, it could later explode into a devastating conflagration. But Singh was unyielding in his resistance. Finally, the district magistrate authorized an unprecedented late-night raid on the village to forcibly vaccinate Mohan Singh and his followers.

As a civil libertarian, Brilliant had serious qualms about resorting to such strong-arm tactics. Other groups in India resisted vaccination for various reasons, including religious beliefs, fear, or misinformation. In one case, an Indian health worker had extorted money from gullible villagers by claiming that Soviet smallpox vaccine would turn them into Communists unless they paid him not to administer it. Whatever the reason, vaccinating people by force was nearly always counterproductive, triggering riots and stiffening resistance. A far better approach was gentle persuasion, usually by gaining the support of a village elder, teacher, or spiritual leader. Because of Mohan Singh's stubborn refusal to compromise, however, smallpox was continuing to spread. Brilliant's commitment to fighting the disease finally prevailed over his respect for Singh's religious beliefs.

The raid on Mohan Singh's home took place at at night when most of the villagers were asleep. According to a published account of the operation by Brilliant and his wife, Girija,

> In the middle of the gentle Indian night, an intruder burst through the bamboo door of the simple adobe hut. He was a government vaccinator, under orders to break resistance against smallpox vaccination. Lakshmi Singh awoke screaming and scrambled to hide herself. Her husband leaped out of bed, grabbed an ax, and chased the intruder into the courtyard. Outside, a squad of doctors and policemen quickly overpowered Mohan Singh. The instant he was pinned to the ground, a second vaccinator jabbed smallpox vaccine into his arm.
>
> Mohan Singh . . . squirmed away from the needle, causing the vaccination site to bleed. The government team held him until they had injected enough vaccine; then they seized his wife. Pausing only to suck out some vaccine, Mohan Singh pulled a bamboo pole from the roof and attacked the strangers holding his wife. While two policemen rebuffed him, the rest of the team overpowered the entire family and vaccinated each in turn. Lakshmi Singh bit deep into one doctor's hand, but to no avail.

When it was over, our vaccination team gathered in the courtyard. Mohan Singh and his exhausted family stood by the broken door of their house. We faced each other silently across a cultural barrier, neither side knowing what to do next. . . .

Mohan Singh surveyed his disordered household and reflected. For a moment or two he hesitated. Then he strode to his small vegetable plot and stopped to pluck the single ripe cucumber left on the vine. Following the hospitality creed of his tribe, he walked over to the puzzled young Indian doctor whom his wife had bitten and handed him the cucumber.

I had stood in the shadows trying to fathom the meaning of this strange encounter. Now I reached out to Zafar Hussain, a Muslim paramedical assistant assigned to me by the Indian government as guide and translator. What on earth was the cucumber for? Speaking in Hindi, Zafar passed my question along to one of the vaccinators, a Westernized Ho youth, who challenged Mohan Singh in the staccato rhythm of the tonal Ho language.

With great dignity, Mohan Singh stood ramrod straight. The whole village was awake now, people standing around the courtyard stage as the rising sun illuminated our unfolding drama. Measuring his words carefully, Mohan Singh began:

"My *dharma* [religious duty] is to surrender to God's will. Only God can decide who gets sickness and who does not. . . . Daily you have come and told me that it is *your* dharma to prevent this disease with your needles. We have sent you away. Tonight you have broken my door and used force. You say you act in accordance with your duty. I have acted in accordance with mine. It is over. God will decide. Now I find that you are guests in my house. It is my duty to feed guests. I have little to offer at this time. Except this cucumber."

Once Mohan Singh had spoken, the five hundred villagers agreed to be vaccinated without further resistance. With the containment of the outbreak in Mohan Singh's village and the onset of the monsoon season, the incidence of smallpox in southern Bihar gradually subsided. Active searches in the towns and villages of the state continued to turn up smallpox cases, but they were located within a shrinking geographical area.

In late 1974, the active search program was expanded to cover all of India. During a six-day period each month, health workers visited every one of the country's 100 million households. Supervised by about fifty international advisers and an equal number of Indian officials, some thirty-three thousand district health personnel and more than 100,000 additional field workers conducted house-to-house searches in a total of 575,721 villages and 2,641 cities. The scale of the effort was colossal in every way: Grasset estimated that each nationwide search consumed about eight tons of paper forms.

As the number of smallpox cases in India diminished, the WHO progressively refined its surveillance-containment strategy. In January 1975, the country team launched "Operation Smallpox Zero," which involved a further intensification of active search operations in the districts that were still infected. Whereas health workers had previously checked only a sample of houses in a village or urban neighborhood, they now went to every house and searched the local markets, fairs, and hospitals. In addition, the size of the containment teams was expanded to vaccinate all persons living within a ten-mile radius of each infected village.

The WHO and the Indian government also sought to increase public participation in the campaign by offering a cash reward of 100 rupees ($12) to any health worker or villager who reported a fresh case of smallpox. Small aircraft flew over remote jungle villages and dropped leaflets informing villagers about the reward program. As the incidence of smallpox declined, the size of the reward was gradually increased until it reached the maximum credible sum of 1,000 rupees, equivalent to several months' wages.

This incentive program had the desired effect: In 1975, 11 percent of small-pox outbreaks were reported by the general public, compared with only 2.6 percent in 1973.

Dozens of times during the Indian campaign, its ultimate success had seemed to hang by a thread. But suddenly, the long struggle was over. In May 1975, health workers found the last case of smallpox in India: a thirty-year-old homeless Bangladeshi beggar named Saiban Bibi, who had developed a fever and rash while living on the platform of a railway station in the Indian state of Assam. In early July, she was discharged from the hospital and returned to Bangladesh. Thanks to rapid and effective containment operations, no secondary cases occurred.

Less than two years after the Indian campaign had begun and fourteen months after its darkest moment—when nearly fifty thousand cases had been reported in Bihar in a single month—children throughout the vast nation were finally safe from the burning fever and painful pustules of smallpox. Surprisingly, however, villagers continued to visit shrines to Shitala Mata to ward off other rash-causing diseases. The WHO country team had hoped to put the goddess out of business, but she still had plenty of measles, scabies, and chickenpox to deal with.

Because August 15, 1975, was India's Independence Day, Prime Minister Indira Gandhi decided to declare it "Independence from Smallpox Day." The Indian government invited WHO Director-General Halfdan Mahler to come to New Delhi for the celebration. Mahler, a Dane, had replaced Marcolino Candau in July 1973 and, in contrast to his predecessor, Mahler was a strong supporter of the smallpox eradication program. Still, it was with some trepidation that D. A. Henderson agreed to the Indian government's request. He feared that a small focus of infection might still exist somewhere in India because of its vast population and the fact that many states were poorly administered. Contributing to his sense of unease, a few hidden cases of smallpox had been discovered months after "zero pox" had been declared in Indonesia, Brazil, and Nigeria, causing the WHO considerable embarrassment.

Despite Henderson's anxieties, the ceremony went off without a hitch. The Indian Minister of Health, Dr. Karan Singh, presented Dr. Mahler with an eight-ton bronze statue of the Hindu god Shiva for display at the WHO headquarters building in Geneva. After the ceremony, Henderson and Mahler planned to fly to Bangladesh, the last country in South Asia where smallpox was still widespread.

The WHO had been struggling to eliminate smallpox from Bangladesh ever since January 1972, when large numbers of infected refugees returning from India had reintroduced the disease on a large scale. In 1973, more than thirty-two thousand cases of smallpox had been reported. The following summer, the WHO had seemed on the verge of containing the epidemic when heavy monsoon rains caused severe floods in smallpox-infected areas, destroying villages and crops and forcing hungry villagers to migrate to the shantytowns of Dacca and other major cities, spreading the disease widely. Then, in January 1975, Bangladeshi president Sheikh Mujibur Rahman launched a slum clearance program that uprooted an estimated 50,000 to 100,000 people, triggering a new string of epidemics. Increasingly, Bangladesh had looked like the place where the global eradication program might well come to grief.

With India approaching zero pox, Henderson had met with Bangladeshi government officials and warned them that unless they got their epidemic under control, the entire country might have to be placed under international quarantine. In February 1975, President Rahman had issued a presidential decree declaring the smallpox epidemic a national emergency. With financial assistance from Sweden, the WHO had launched an intensified active search campaign in which health workers visited 87 percent of the 15 million homes in the entire country at least once. By July 1975, the number of infected villages had fallen below 150, and zero pox was projected for the fall.

That was the situation on the evening of August 15, 1975, when Henderson and Mahler were preparing to leave the Independence Day celebration in New Delhi for the airport and the short flight to Dacca. Suddenly, they received a news flash that the Bangladeshi military had staged a coup and that President Rahman and several members of his household had been assassinated. The coup leaders had closed the airport, sealed the borders, and imposed a news blackout.

In New Delhi, rumors of civil unrest and refugee flows were rife. It seemed likely that Bangladesh would soon be engulfed in civil war, potentially dealing a fatal blow to the smallpox eradication campaign. Henderson feared a repeat of the massive migration of smallpox-infected refugees that had occurred in late 1971 and early 1972, only this time bringing smallpox back to India. To help prevent this nightmare, the WHO deployed large numbers of Indian health workers to patrol the border with Bangladesh. Surveillance teams searched villages within a sixteen-kilometer-wide belt along the Indian side of the frontier, as well as districts in the city of Calcutta. Fortunately, the border area remained quiet and no importations of smallpox occurred.

After a few tense weeks, the crisis passed and the new Bangladeshi military government reopened the borders. The WHO surveillance workers, who had kept a low profile out in the countryside, immediately went back to work. In mid-September, the country team believed that it had found the last smallpox-infected village in Bangladesh, outside the city of Chittagong. This area was under the command of an army general who initially denied the WHO permission to vaccinate. French epidemiologist Daniel Tarantola finally persuaded the general to relent, and the outbreak was snuffed out.

After two months had passed without fresh cases, the WHO held a press conference on November 14, 1975, to announce that Bangladesh was smallpox-free. Major newspapers, including the *New York Times,* ran the story, and the WHO country staff converged on Dacca for a celebration. The next morning, however, a telex arrived reporting a suspected case of active smallpox in a remote village called Kuralia on Bhola Island. Located far to the south at the mouth of the Ganges River, Bhola Island was a swampy low-land of rice paddies, sand, and palm trees with a population of 960,000. In 1970, a cyclone and a tidal wave had swept away scores of villages, leaving the local health services in disarray.

A team of WHO epidemiologists set out immediately for Kuralia, expecting to find a misdiagnosed case of chickenpox. After a twenty-four-hour journey by speedboat, steamer, jeep, motorcycle, and on foot, they arrived at the village. There they found a two-and-a-half-year-old girl named Rahima Banu recovering from a clear case of smallpox, with a pockmarked face and a few scabs still clinging to her legs. Two days later, about two hundred more epidemiologists and field workers arrived at Bhola Island by boat from other parts of the country, bringing with them additional supplies of vaccine, food, vehicles, and equipment.

Rahima and her family were quarantined at home and containment teams vaccinated more than eighteen thousand people living within a 1.5-mile radius. Meanwhile, surveillance teams began making repeated sweeps of villages throughout a five-mile radius, searching for smallpox cases house by house and visiting markets, schools, and tea shops. The WHO also publicized a large cash reward for reporting a confirmed case of smallpox. Although thousands of reports flowed in, all of them turned out to be chickenpox or measles. Rahima Banu was the last victim of naturally occurring variola major anywhere in the world. Three years and eleven months after smallpox had been reintroduced from India into Bangladesh, South Asia was finally free of the disease and the long-sought goal of global eradication was in sight. Only one country in the world still harbored endemic smallpox: Ethiopia, on the Horn of Africa.

Until the 1920s, variola major had been the predominant form of smallpox in Ethiopia, but by the early 1970s, variola minor (alastrim) was the only variety found there. Because the disease produced a sparse rash and had a mortality rate of only 2 percent in adults and 12 percent in infants under the age of one, the Ethiopian government did not consider small-pox eradication to be a high priority.

Ethiopia also posed immense logistical and cultural challenges. Twice the size of Texas, it was inhabited by 25 million people at the time, making it the third most populous country in Africa. More than 90 percent of the population was rural, and a patchwork of ethnic groups spoke more than seventy languages and two hundred dialects. Communications were rudimentary and half the population lived more than three days' walk from an accessible road. Lacking modern health care, people in rural parts of Ethiopia relied on traditional shamanistic ceremonies to ward off smallpox. The centuries-old practice of variolation was also widespread, although it was far riskier than vaccination and often spawned additional epidemics.

Beginning in 1973, the WHO focused its efforts in the Ethiopian highlands, a region of high mountains and deep gorges crisscrossed by rivers that became impassable during the six-month rainy season. Because the area had few paved roads, two-person mobile surveillance teams traveled by mule and on foot to reach isolated villages, search for smallpox cases, and vaccinate. In November 1974, the WHO began to use helicopters to ferry the teams to more remote areas, after which they walked for seven to ten days to a prearranged pickup point.

The two main ethnic groups in the central highlands, the Amhara and the Dorsey, strongly resisted vaccination for cultural reasons. Some of the mobile surveillance teams were chased out of villages with spears and dogs, but others remained in an area for several days, meeting with Coptic priests, tribal chieftains, and clan elders in an effort to discourage variolation and gain the cooperation of the local population. Weeks of cajoling could be required to vaccinate half the inhabitants of an infected village. Some of the natives accepted vaccination only if it was done on the wrist, the traditional site of variolation, or if other medications were given first. They often demanded treatment for unrelated health problems, or brought individuals who had been blinded by smallpox and asked the doctors to restore their sight. In an effort to gain the villagers' trust and encourage their compliance with the immunization program, the WHO teams provided some general medical care and referred people with serious illnesses to the nearest hospital.

Sporadic fighting between Ethiopian government forces and several ethnic groups seeking independent homelands also hampered the vaccination campaign. Parts of the central highlands were war zones, and the mobile surveillance teams were often at risk. Several frightening incidents occurred: A WHO helicopter was destroyed by a hand grenade, a second damaged by rifle fire, and a third captured and the Canadian pilot held for ransom; and two local health workers were killed by bandits. Despite the dangers, however, the work went on. In March 1976, vaccinators with armed escorts contained the last outbreaks of smallpox in the central highlands.

From then on, the WHO country team concentrated its efforts on the Ogaden Desert, a vast expanse of dry savannah straddling the border between Ethiopia and Somalia. This region was inhabited mainly by nomadic herders who traveled in extended family groups of thirty to forty persons from one encampment to the next, following the seasonal rains. The encampments were little more than clearings cut in the brush and surrounded by a crude fence of thorny branches to keep out wild animals. Within these clearings, the nomads erected small dome-shaped huts consisting of curved sticks covered with animal hides. They grazed their herds of goats and camels on the sparse vegetation of shrubs and acacia trees, subsisting on meat, camel milk, and a few vegetables purchased in local markets. When the animals ran out of fodder after a week or so, the nomads dismantled their huts, packed them onto camels, and moved on.

Because the frequent movements and changing encampments of the Ogaden nomads frustrated efforts to trace the chains of smallpox transmission and vaccinate systematically, the WHO country team mapped all the water holes in the area. They then hired more than two hundred local health workers to crisscross the desert, searching for smallpox cases. After a long and persistent effort, the last outbreak in the Ogaden was discovered at a nomadic encampment called Dimo in July 1976 and was fully contained by the following month.

Smallpox now appeared to have been eliminated from Ethiopia, and perhaps from the world. The WHO team members celebrated their achievement by throwing a party at their main Ogaden camp in the town of Gode. D. A. Henderson flew in from Geneva for the occasion, and a reporter and photographer from *National Geographic* arrived to document the event. In the midst of their rustic meal, a radio call reported a possible case of smallpox in a village in the central highlands, putting a damper on the celebration. Early the next morning, Henderson and Brazilian epidemiologist Ciro de Quadros flew north by helicopter to examine the patient, who turned out to have chickenpox.

For seven weeks, no new cases of smallpox were reported anywhere in the world. Because the WHO eradication program appeared to be on the verge of reaching its final goal, Henderson accepted an offer to become dean of the school of public health at Johns Hopkins University, and Isao Arita took over as head of the Smallpox Eradication Unit. In September 1976, however, five cases of smallpox turned up unexpectedly in Mogadishu, the capital of Somalia, which bordered Ethiopia on the Horn of Africa. This outbreak, the first in Somalia in fourteen years, came as a bitter disappointment for the WHO staff. The mood turned even more grim when it became clear that the epidemic had spread. A nationwide search in March 1977 revealed more than three thousand cases throughout the southern half of Somalia, putting neighboring Kenya and Ethiopia at risk. Henderson also feared that infected Somali Muslims might go on pilgrimage to Saudi Arabia, reintroducing smallpox into the Middle East.

On May 18, the Somali government declared a national emergency and appealed for international assistance. Several countries made cash donations and in-kind contributions totaling more than $400,000, and shipments of vehicles and equipment began arriving at Mogadishu Airport. The epidemic peaked at 1,388 cases in June 1977, by which time twenty-three international advisers and more than three thousand Somali health personnel were working to contain it. Because the outbreaks were surprisingly widespread, it became clear that surveillance-containment alone would not stop transmission, and countrywide vaccination programs were implemented. Then, in July, the Somali Army invaded the Ogaden Desert in a bid to wrest the contested region from Ethiopian control. For nearly two months, WHO advisers were denied access to the war zone. In response, the WHO country team established five base camps near the disputed Somali-Ethiopian border and recruited local nomads to continue searches on foot throughout the Ogaden, avoiding areas of active combat. No further cases of smallpox were found.

In mid-August 1977, what appeared to be the last outbreak of smallpox in southern Somalia struck a nomadic group of about twenty families with a total of 109 members. Over the next two months, as the nomads wandered across the dry savannah, eight children came down with the disease. On the evening of October 12, at a nomadic encampment ten miles north of the village of Kurtunwarey, the local authorities found two children with active smallpox. One of the children, a six-year-old girl named Habiba Nur Ali, was critically ill, while her younger brother had a rash in the early stages. The officials took the children and their mother to a local health post, which arranged for a ministry of health vehicle to drive them to an isolation camp near Merca, a town of twenty-six thousand people on the Indian Ocean coast.

On the way to the isolation camp, the driver stopped at Merca Hospital to ask for directions. Ali Maow Maalin, the twenty-three-year-old hospital cook, climbed into the back of the Land Rover and sat next to the two sick children for less than ten minutes as he guided the driver to his destination. Tragically, Habiba died in the isolation camp two days later, the world's last known fatality from endemic smallpox. Although a follow-up investigation ensued, Ali Maow Maalin was not identified as a possible contact.

Meanwhile, Peter Carrasco, an American epidemiologist working for the WHO in Somalia, learned of the smallpox outbreak near Kurtunwarey and arrived on October 13 with several colleagues just as the nomads were preparing to move out. A few small children still had drying scabs that contained the live virus. Carrasco got permission from their mothers to pull off the scabs, which he burned in a wood fire, and he then vaccinated all possible contacts. On learning that some of the young nomad men had left with their camels a few days earlier to search for water, Carrasco and his fellow epidemiologists decided to follow them into the desert.

Because the heavy summer rains had washed out roads and river crossings, severely impeding vehicular travel, the international advisers split up into three teams and headed on foot toward the rain clouds on the horizon, using hired camels and donkeys to carry their tents, food, and supplies. Over the next two weeks, the teams of epidemiologists made a broad swath across the desert in a classic military pincer movement, following the nomads from one encampment to the next and vaccinating everyone they encountered. On October 18, near a water hole called Edi Shabelli, they finally contained the last outbreak.

The Somali epidemic was not quite over, however. Although Ali Maow Maalin had worked temporarily as a smallpox vaccinator, he had not been successfully vaccinated himself, and his brief exposure to the two sick children had been enough to infect him. On October 22, he developed a fever at work and went home to recover. Three days later, he was admitted to the medical ward at Merca Hospital with a sparse skin rash and was discharged the next day with a presumptive diagnosis of chickenpox. Bedridden in his rented room near the hospital, Maalin suspected that he might have smallpox but did not tell the authorities for fear of being quarantined. Finally a coworker at the hospital, seeking a reward, reported him and he was placed in the isolation camp, where he recovered without complications and was discharged at the end of November.

Because Maalin had been visited by numerous friends and relatives during his illness, the WHO called in epidemiologists from throughout

Somalia to help contain the potential outbreak. They closed Merca Hospital to new admissions, vaccinated the medical staff, and quarantined patients in the hospital with a twenty-four-hour police guard. Ninety-one people who had been in face-to-face contact with Maalin were vaccinated and placed under observation, but none developed smallpox. Containment teams also vaccinated everyone in the fifty houses surrounding Maalin's home and later throughout the entire city ward, and set up checkpoints on the road and footpaths leading into Merca so that everyone entering or leaving the city could be stopped and vaccinated. In all, more than fifty-four thousand people were vaccinated during the two-week period from October 31 to November 14, 1977.

Meanwhile, the war in the Ogaden continued. Because of the fighting and heavy rains that washed out roads, it was not until the spring of 1978 that the WHO was able to resume active searches throughout Somalia and confirm that the smallpox epidemic was indeed over. On April 17, 1978, the WHO field office in Nairobi, Kenya, sent a brief but historic telegram to Geneva. It read: "Search complete. No cases discovered. Ali Maow Maalin is the world's last known smallpox case." The Somali cook was the endpoint of a continuous chain of human-to-human transmission extending back thousands of years.

SEVEN

REALM OF THE FINAL INCH

With the last case of smallpox in Somalia, the final phase of the global campaign began. To certify that the disease had truly been eradicated from the world, the WHO Smallpox Eradication Unit decided to require two years of active surveillance after the last reported case in each formerly endemic country to make sure that no hidden pockets of smallpox remained. The rationale for this length of time was that smallpox, in order to persist, had to pass in an uninterrupted chain of infection from one person to another. Each infected individual developed a distinguishing rash. Since the virus could not lie dormant anywhere, be carried by animals, or cause silent infections in humans, the persistence of smallpox in any area would eventually become known. Experience had shown that of all the countries that were declared smallpox-free, the longest time that a pocket of the disease had remained undetected was four months. That situation had occurred in Indonesia, where local health officials had deliberately concealed smallpox cases. Thus, requiring two years of surveillance in each formerly endemic country would provide a high level of confidence that global eradication had in fact been achieved.

The certification process began in August 1973 in South America and ended in October 1979 in Somalia. Mobile teams conducted systematic house-to-house searches, and "rumor registers" were established to follow up on all reports of suspected cases. An important element of the certification process was the offer of a substantial cash reward to anyone who reported a fresh case of smallpox, although this policy generated a torrent of false leads that had to be run down. Because smallpox and severe chickenpox could sometimes be confused clinically, the WHO reference laboratories at the CDC in Atlanta and the Research Institute for Viral Preparations in Moscow analyzed hundreds of tissue specimens from suspected cases, all of which proved negative.

Maintaining a high level of vigilance throughout the certification period was one of the toughest challenges of the entire campaign. Each formerly endemic country was understandably eager to transfer money and manpower to other pressing health problems and was not enthusiastic about sustaining two more years of intensive surveillance to confirm that smallpox elimination had been achieved. For D. A. Henderson, this phase was akin to the "realm of the final inch" described in Alexander Solzhenitsyn's novel *The First Circle*—the painful search for perfection in completing a complex task.

The two years of active surveillance in each formerly endemic country were followed by visits from expert committees known as International Commissions for the Certification of Smallpox Eradication, twenty-two of which operated between 1973 and 1979. Some of the certification teams visited a single large country, such as Indonesia or Bangladesh, while others were assigned three or four smaller countries that had not reported cases of smallpox for many years. Each commission carefully reviewed the documentary evidence provided by the host governments and conducted its own evaluations in the field. In April 1974, Indonesian Minister of Health Julie Sulianti Saroso set the tone for all subsequent visits by certification teams when she said, "Go anywhere you wish, ask anyone any question, examine any document. I am confident that you will find that smallpox has been eliminated from Indonesia."

The twenty-two international commissions reported directly to the Global Commission for the Certification of Smallpox Eradication, which WHO Director-General Mahler established in 1978. Consisting of twenty-one experts from nineteen countries, the Global Commission was chaired by Frank Fenner, a leading poxvirologist from Australia, and D. A. Henderson served as a member. The mandate of the Global Commission was to review the certification data from all countries where smallpox had been endemic in 1967 or later and either pronounce itself satisfied that global eradication had been achieved or suggest what additional actions or studies were warranted.

The WHO also asked the 121 countries that had become smallpox-free before 1967 to provide formal documentation, but the People's Republic of China did not comply. Because the PRC was not then a member of the WHO, Beijing refused to provide data corroborating its claim of having eliminated smallpox in 1965. When the Global Commission decided that it needed more definitive evidence, the WHO's only recourse initially was to ask travelers who visited China or Hong Kong to look for children with pockmarked faces, of whom none were seen. This anecdotal information was not considered conclusive, however. Finally, through the intercession of the Australian government, a WHO delegation consisting of Global Commission chairman Frank Fenner and epidemiologist Joel Breman visited China in 1979. They focused their attention on densely populated areas that were most likely to have harbored smallpox at some time during the preceding fourteen years of claimed freedom from the disease. After traveling around the country for five weeks and reviewing the extensive Chinese documentation, they satisfied themselves that zero pox had been achieved.

Members of the Global Commission did not attempt to visit Tibet or North Korea. Nevertheless, because smallpox could persist in a population only by being transmitted continually from person to person, the longer the amount of time that had elapsed since the disease had been present in a given country, the greater the confidence that it was no longer circulating. No cases had been reported in the Koreas since 1954 and none was rumored. Similarly, Tibet was a sparsely populated country with no evidence of smallpox for many years. Thus, based on all of the available evidence, the Global Commission was able to reach a unanimous judgment in December 1979 that smallpox had in fact been eradicated from the planet.

Although smallpox had now been vanquished as a human disease, frozen samples of variola virus were still stored in scientific and clinical laboratories around the world. In response to a survey conducted in 1975 by the WHO, a total of seventy-five laboratories reported holding scabs or viral isolates from smallpox patients: twenty-nine labs in Europe, eighteen in the Americas, thirteen in Southeast Asia, seven in the Western Pacific, five in Africa, and three in the Eastern Mediterranean. The total number of sites with stocks of variola virus was fairly small for several reasons: the fact that smallpox had usually been diagnosed on the basis of clinical symptoms; the existence of the two WHO reference laboratories in Atlanta and Moscow, which had done the lion's share of diagnostic work during the eradication campaign; the preference of most virologists for studying other poxviruses because of the lack of an animal model of smallpox; and the difficulty in most developing countries of obtaining the uncontaminated chicken eggs needed to culture variola virus.

Despite the limited number of labs retaining the live virus, the possibility of an inadvertent release was still worrisome. Indeed, two accidents during the 1970s had sparked concern about the hazards of smallpox research. The first incident had occurred on February 28, 1973, when Ann Algeo, a twenty-three-year-old technician at the London School of Hygiene and Tropical Medicine, visited the poxvirus laboratory headed by Dr. Charles Rondle and observed the harvesting of egg membranes infected with variola major. At that time, virologists cultured and manipulated the virus on open lab benches, relying for protection on frequent vaccination, careful laboratory technique, and the sterilization of cultures and glassware. Although Algeo had been vaccinated against smallpox as a child, a failed revaccination had left her susceptible to infection. She fell ill on March 11 and was hospitalized in the general ward at St. Mary's Hospital in London. Nine days later, she was diagnosed with smallpox and transferred to an isolation hospital in Dartford, Kent, where she recovered. During Algeo's hospitalization in London, however, she had unknowingly infected the son and daughter-in-law of the patient in the next bed when they came to visit. At the end of March, the young couple developed smallpox and subsequently died.

In response to this tragedy, a committee of inquiry established by the British government determined that Dr. Rondle's lab had conducted smallpox research without adequate physical containment. The committee recommended a code of safety for all future work with variola, including formalized procedures for the revaccination of researchers and the use of sealed biosafety cabinets to isolate the virus from the surrounding air, particularly when laboratory manipulations were likely to generate infectious aerosols. The British government also created a standing expert committee called the Dangerous Pathogens Advisory Group to help ensure laboratory safety. Accidents involving smallpox were not limited to England, however. In 1975, a chimpanzee infected with variola escaped from the poxvirus laboratory headed by Albert Herrlich at the Institute for Comparative Tropical Medicine of the University of Munich. The chimp remained at large for several days, swinging from tree to tree in a nearby park, before it was finally captured.

Concerned about the risk of an accidental release or theft of variola virus from a laboratory after smallpox had been eradicated, the WHO sought to limit the number of research centers that retained stocks of the virus. The Smallpox Eradication Unit prepared a resolution for the World Health Assembly in May 1976 requesting all institutions that no longer did research on variola virus to destroy their stocks or transfer them to a small number of designated "WHO collaborating centers." Although the WHO had no power of enforcement, this effort was quite successful. By the end of 1977, eighteen laboratories in the world reported that they still retained stocks of live variola virus, down from the original number of seventy-five.

One of the places in Europe that continued to do smallpox research was the medical school at the University of Birmingham, England's second-largest city. Professor Henry S. Bedson, aged forty-eight, directed the pox-virus laboratory there. Balding but with an earnest, boyish face, he had a dry sense of humor and was well respected by his scientific peers. Because Bedson had been promoted to head of the Medical Microbiology Department in September 1976, his increased administrative duties left him little time to spend in the laboratory. He therefore delegated much of the experimental work to a doctoral student. Also working in Bedson's lab were a senior technician who had been at Birmingham for eleven years, and a trainee technician with only nine months' experience.

Research on smallpox at Birmingham was restricted to a tiny room, eight feet square, adjacent to a larger laboratory reserved for work with animal poxviruses that did not normally infect humans. In early 1977, Bedson began investigating "whitepox," a newly discovered species of poxvirus that researchers in Moscow and the Netherlands claimed to have isolated from wild monkeys and African rodents. When this virus was grown on egg membrane, it produced white pocks identical to those of variola, in contrast to the red, hemorrhagic lesions produced by monkeypox virus. Also like variola, whitepox did not grow in rabbits, guinea pigs, or mice. The discovery of an animal virus that so closely resembled the causative agent of smallpox had disturbing implications. Although considerable epidemiological evidence argued against the existence of an animal reservoir for variola, it was possible that whitepox virus was a close relative that might be capable of infecting humans. Indeed, because whitepox could not be distinguished from variola in standard diagnostic tests, cases of a human disease attributable to whitepox might already have occurred and been misdiagnosed as smallpox.

Bedson sought to identify some distinguishing characteristics between the whitepox and variola viruses that might help unravel the mystery. He suspected that the two viruses differed in some of their protein constituents, such as the enzymes involved in viral replication or the structural building blocks of the viral coat. Because these proteins were present only in minute amounts, analyzing them biochemically required growing large quantities of each virus in fertilized chicken eggs. Accordingly, the doctoral student and the two technicians in Bedson's lab began to cultivate whitepox and several different strains of variola and purify the viral proteins. To harvest the viruses, they spun crude extracts in a low-speed centrifuge, a process known to generate infectious aerosols.

Meanwhile, Professor Bedson experienced a depressing setback. The World Health Assembly had passed a resolution in 1976 urging that all stocks of variola virus be transferred to a small number of designated WHO collaborating centers, where further research on smallpox might proceed. Bedson had asked to have his laboratory designated an official collaborating center, but senior WHO officials turned down his request in September 1977. This decision came as a deep disappointment to Bedson at a time when his whitepox experiments were yielding promising results.

On May 4, 1978, a three-man WHO inspection team visited Birmingham. Earlier, an expert group convened by the Smallpox Eradication Unit had drawn up new biosafety guidelines for smallpox research, and each laboratory that worked with the virus was being evaluated as to its success in meeting those guidelines. On visiting Bedson's smallpox laboratory, the WHO inspectors found that the equipment and procedures for physical containment were unsatisfactory and recommended that the lab either be brought up to the new standards or closed at the earliest possible date. Although the WHO safety inspectors could point out deficiencies and make recommendations, they did not have the power to close down an unsafe facility; that was the responsibility of each national government.

Bedson could not afford to make the recommended improvements. In a letter to the WHO on June 2, he claimed that his research posed negligible risks, citing the "progressive decline in the scale and diversity of our operations" since 1973 and the "marked increase in the level of physical containment which has been introduced." Because Bedson was widely regarded as a skilled and experienced virologist, the British authorities accepted his proposal to finish all work with variola and whitepox by the end of 1978. In fact, the assurances in Bedson's letter to the WHO were false, and he had also been less than forthright with the safety inspectors about some of his laboratory's more hazardous operations. Far from reducing the scale of his operations, he had just received delivery from his friend and colleague, Keith Dumbell, of twenty-two strains of variola virus, in addition to the fourteen strains he already held, to test the ability of a biochemical technique to distinguish among them. Moreover, the workload in the smallpox lab at Birmingham soon increased tenfold because of the urgency of completing the whitepox experiments by the December 1978 deadline.

On August 11, 1978, Janet Parker, a forty-year-old medical photographer in the Department of Anatomy at Birmingham Medical School, became ill with fever, headache, and muscular aches and went home to recover. She worked on the floor directly above Bedson's animal poxvirus lab and often made calls from a small telephone cabinet next to the darkroom. Seventeen days earlier, on July 25, she had spent most of the day on the phone ordering equipment and materials from photographic suppliers.

Believing she had a cold or flu, Parker stayed home from work. The next day she went to her personal physician, who prescribed an antibiotic. Her symptoms worsened, however, and red spots appeared on her face, limbs, and trunk. Two days later, her doctor's partner assessed the rash to be an allergic reaction to the antibiotic and stopped the medication. On August

21, when the red spots turned into blisters, Parker moved to her parents' home, where she remained bedridden. Three days later, she was examined by her parents' doctor, who immediately suspected smallpox and had her admitted that afternoon to East Birmingham Hospital and placed in strict isolation. Although Parker had been vaccinated against smallpox in 1966, she had not been revaccinated since then because her work at the medical school did not involve direct contact with the virus.

On the evening of August 24, a swab specimen of Parker's rash was delivered to the University of Birmingham poxvirus laboratory, where Professor Bedson prepared it for electron microscopy. Working late into the night, he fixed and stained the material, which he mounted on a round copper grid about three millimeters in diameter. Using tweezers, he placed the grid in the sample chamber of the electron microscope. He then switched on the vacuum pump, evacuating the air from the sample chamber and the stainless steel cylinder above it, so that a narrow beam of electrons could be focused down through the specimen.

Sitting in the dimly lit room in front of the gleaming instrument, Bedson stared anxiously at the fluorescent screen. When he focused the electron beam, the minute divisions of the copper grid appeared on the screen as thick black bars. Between them, magnified some 40,000 times, were small biscuit-shaped objects that he immediately recognized as poxvirus particles. Although Bedson could not identify the species of poxvirus from its external appearance alone, it was clear from the circumstances that Parker was infected with variola major—a disease that supposedly had been eradicated from the planet. The fact that Bedson was the one to identify the virus that had escaped from his own laboratory was one of the darker ironies of the case.

At 10:00 P.M. that night, Janet Parker was transferred to the Catherine-de-Barnes Isolation Hospital near Birmingham. When her diagnosis became known, the University of Birmingham immediately closed Bedson's poxvirus laboratory and sealed off the entire wing of the medical school building as a precaution. On August 27, Keith Dumbell at St. Mary's Hospital Medical School in London performed a genetic analysis of the virus isolated from Parker. He determined that she had been infected with the "Abid" strain of variola major, named after a young Pakistani boy from whom the virus had first been cultured. Abid was one of the smallpox strains that had been cultivated since May in Bedson's laboratory.

Meanwhile, the Birmingham health authorities identified 341 people who had come into close or casual contact with Parker and might have become infected. These individuals included the members of her immediate family and other relatives, her colleagues at the medical school, and the patients at East Birmingham Hospital where she had been hospitalized initially. All of these presumptive contacts were notified, vaccinated, quarantined at home, and checked for symptoms on a regular basis. The authorities even tracked down a twenty-year-old British woman who worked with Parker and had traveled to the United States on August 18 to visit a farm in North Dakota.

On August 30, the British government's Department of Health and Social Security appointed Reginald A. Shooter, a professor of microbiology and chairman of the Dangerous Pathogens Advisory Group, to conduct an official investigation of the Birmingham incident. By this time, criticism of Professor Bedson was intensifying. Blaring headlines in the Fleet Street tabloids attacked him personally, and the trade union to which Parker belonged demanded angrily to know why the university had allowed the smallpox laboratory to remain open despite its failure to meet international safety standards. Disconsolate over Parker's illness, Bedson took much of the criticism to heart. He was plagued by feelings of guilt over the accident and tormented by the newspaper articles vilifying him. His scientific reputation had been tarnished, perhaps irrevocably. On August 31, he was heard to remark, "It seems that whatever the result of inquiries into the way the member of staff became infected with smallpox, our project will have to end."

The following day, Bedson told his wife Ann that he was going out to do some gardening. She was talking on the telephone and watched from the window as he disappeared into the potting shed next to the house. When he failed to emerge after several minutes, she went down to investigate. Entering the potting shed, she found her husband lying on the floor in a pool of blood, his throat slashed open with a knife. He had left a suicide note that read: "I am sorry to have misplaced the trust which so many of my friends and colleagues have placed in me and my work. . . . I realize this act is the least sensible I have done, but it may, I hope in the end, allow [my family] to get some peace." An ambulance rushed the unconscious Bedson to Birmingham Accident Hospital, where he arrived in critical condition and was placed on a respirator. Five days later, on September 6, doctors determined that he was brain-dead and switched off the life-support machine.

Meanwhile, Janet Parker's father had developed a fever that was probably a side effect of smallpox vaccination, but he was admitted to the isolation hospital as a precaution. On September 5, he died suddenly of a heart attack. Two days later, Parker's seventy-year-old mother became ill while in quarantine at her home. Although she had been vaccinated and treated with an immune serum, she developed a sparse rash of red spots that proved to be a mild case of smallpox. The tragedy was not yet over. Janet Parker's condition continued to deteriorate, and she died on September 11 from complications of smallpox, including kidney failure and a bacterial infection.

Eight other people, mainly close contacts of Janet Parker, came down with fever and were admitted to the hospital as a precaution, but their symptoms were all assessed to be side effects of vaccination. Parker's mother recovered uneventfully and she was released from isolation on September 22. When no additional cases appeared, the emergency was declared over. Only luck had prevented a major outbreak in Birmingham of a particularly virulent strain of smallpox. Because Great Britain did not require routine vaccination against the disease, much of the city's adult population was susceptible to infection.

The investigative committee under Professor Shooter found much to criticize about Bedson's laboratory, including clear evidence of carelessness and failure to follow basic safety precautions. With respect to the source of Parker's illness, the committee concluded that the most likely route of infection was through the air. Large quantities of variola virus had been grown and processed in Bedson's lab under hazardous conditions, and some of the manipulations had been done inside a safety cabinet that was later shown to be defective. The Shooter committee hypothesized that aerosolized variola virus had escaped from the smallpox lab and floated into the adjacent animal poxvirus lab. From there, the airborne virus had apparently passed through gaps in an ill-fitting inspection panel on a service duct carrying cables and wires to the telephone room on the floor above, a distance of about eight feet. When seated at the telephone, Parker would have been close to another poorly fitting inspection panel with gaps through which the airborne virus could have leaked. Tests with smoke tracers demonstrated that this path was physically possible, although some experts doubted that normal working conditions in the smallpox laboratory could have generated high enough concentrations of aerosolized virus for infection to occur over such a long distance.

In fact, a case of long-range smallpox transmission had already been reported. In January 1970, a twenty-year-old West German electrician returned from a trip to Karachi, Pakistan, to his hometown of Meschede (near Düsseldorf), where he came down with a high fever, diarrhea, and cough, probably due to a concurrent case of influenza. Initially diagnosed with typhoid fever, he was placed in an isolation room on the ground floor of Meschede General Hospital and remained for three days, with the door usually kept closed. After developing a pustular rash, he was diagnosed with smallpox and transferred to a special quarantine hospital. During his brief stay at Meschede General, the electrician managed to transmit smallpox to seventeen people on all three floors of the building. When investigators placed a smoke-generating device in the patient's room, they found that a nearby stairwell generated strong vertical air currents that had carried the virus to the upper floors. An unusual combination of factors had favored the airborne transmission of smallpox: The patient had a severe case of the disease, his cough was highly efficient at aerosolizing the virus, and the incident took place in winter, when the low relative humidity enhanced the survival time of the virus in the hospital air. The Meschede incident demonstrated the frightening potential of variola to infect at extremely low doses and to travel long distances through the air under special conditions.

Regardless of the exact source of the virus, the Shooter committee concluded that Janet Parker had become infected because of failures in two key lines of defense, vaccination and physical containment. Although Professor Bedson had deliberately misled the university, the WHO, and the British government about the true hazards of his research, the regulatory authorities had also been naïve and careless. The Shooter report concluded: "Only chance and the efficient control measures of the preventive safety authorities prevented a wider spread of infection."

Responding to the political fallout from the Birmingham accident, the British government decided to consolidate all research on smallpox at a single location. In September 1978, the collection of variola virus strains from the University of Birmingham was transferred under police escort to Keith Dumbell's laboratory at St. Mary's Hospital Medical School, which had been designated the sole WHO collaborating center for smallpox in Britain. Because St. Mary's was located in a densely populated part of London, however, the British government soon reconsidered its decision. A year later, in November 1979, the variola virus stocks were again moved for safety reasons. This time the destination was the Centre for Applied Microbiology at Porton Down, a secure government public health laboratory in a rural area near Salisbury.

Meanwhile, the whitepox mystery that had led to Bedson's ill-fated series of experiments was finally solved. The answer was obtained through the application of a powerful new technique called DNA fingerprinting, which was increasingly being employed by police labs to identify criminals from traces of blood, semen, or saliva left at the scene of a crime. This technique made it possible for researchers at St. Mary's Hospital and the CDC to compare the genetic structure of variola virus with that of the whitepox strains that Dutch and Soviet scientists claimed to have isolated from wild monkeys and rodents. Much to their surprise, the researchers found that the DNA fingerprints of the whitepox viruses were identical to those of some well-known strains of variola major. Confronted with this evidence, the Dutch and Soviet groups admitted, red-faced, that their purported discovery of a smallpoxlike virus in animals had been a mistake, the result of inadvertent laboratory contamination. Debunking the whitepox story removed any lingering doubts that smallpox had been eradicated as a clinical disease.

On May 8, 1980, the World Health Assembly in Geneva declared officially that smallpox had been conquered worldwide. The assembly also endorsed the nineteen recommendations contained in the final report of the Global Commission on the Certification of Smallpox Eradication, most notably the call to halt the routine vaccination of civilians against the disease. As insurance against a possible resurgence of smallpox, however, the WHO continued to stockpile 200 million doses of vaccine.

Beyond liberating humanity from a loathsome scourge, smallpox eradication turned out to have been a financial bargain. The total cost of the eleven-year campaign was about $300 million, of which two-thirds had come from the endemic countries themselves in cash or in kind. This investment would now yield substantial dividends, saving an estimated $2 billion a year that had previously been spent on vaccination, inspection, and quarantine. Moreover, in 1974 the World Health Assembly had approved a proposal by the Smallpox Eradication Unit to launch a new vaccination initiative called the WHO Expanded Program on Immunization, which was now delivering measles, polio, and diphtheria-pertussis-tetanus (DPT) vaccines to children around the world.

To commemorate the historic achievement of smallpox eradication, the WHO set aside a special day of solemn ceremonies during the 1980 World

Health Assembly. The official celebration was muted, however. For all its success, smallpox eradication had come under intense criticism because it was a vertically organized program that focused on a single disease, creating a temporary public health infrastructure that dissipated rapidly once the goal had been achieved. The WHO had since changed its priorities and now favored the delivery of basic health services to the masses under the slogan "Health for All by the Year 2000." Moreover, the WHO leadership did not want to lionize the people who had carried out the eradication campaign. Senior WHO bureaucrats resented D. A. Henderson and his colleagues for their brash, freewheeling style and repeated violations of established rules and regulations, which went against the conservative grain of the organization.

Because of the lack of official recognition from the WHO, Henderson decided to establish his own honorary society to pay tribute to the hundreds of international advisers who had labored tirelessly to eradicate smallpox and had since dispersed to public health jobs around the world. In all, 687 individuals from 73 countries had served with the Intensified Smallpox Eradication Program for periods ranging from three months to more than ten years, and about 125 more had participated under bilateral agreements with national agencies such as the U.S. Peace Corps. Henderson designed a lapel pin consisting of a bifurcated needle bent into the shape of a zero, with a clasp to hold it in place, and a certificate inducting the recipient into the "Order of the Bifurcated Needle." Cognizant that the WHO would never officially sanction the award, he had his daughter produce the pins by hand and mailed them out at his own expense. Although Henderson's honorary society was partly tongue-in-cheek, it was motivated by the sincere desire to salute the people who had worked on the eradication campaign, and he was gratified by the positive response.

The Smallpox Eradication Unit had been quite successful at persuading countries to destroy their laboratory samples of variola virus or to transfer them to a small number of designated WHO collaborating centers. After Japan and the Netherlands agreed to ship their strain collections to the CDC in 1981, only three countries other than the United States and the Soviet Union still retained the virus: Britain, China, and South Africa. But this hard-core group of countries strenuously resisted parting with their stocks.

Although a committee of scientists who advised the WHO on smallpox issues favored destroying all remaining specimens of variola virus, they agreed that it would first be desirable to preserve the DNA of some representative strains so that the virus could be identified unequivocally if it ever reappeared in the future. To this end, Keith Dumbell, working with the British strain collection at Porton Down, used a technique known as "gene cloning" to copy fragments of smallpox DNA several millionfold in bacteria. Because the DNA snippets were noninfectious, they could be archived and studied in complete safety.

When Keith Dumbell completed his cloning experiments at Porton Down in September 1982, the residents of neighboring Wiltshire were beginning to oppose smallpox research at the defense laboratory. To comply with the WHO policy of consolidating the virus stocks, the British government decided to transfer its entire collection of variola virus strains to the CDC. In December 1982, the roughly two hundred vials containing viral isolates were packed in dry ice inside Styrofoam containers, sealed in multiple layers of plastic wrap, and placed in the trunks of two cars for the drive to Heathrow Airport. Long sections of the motorway were cordoned off to prevent terrorist activity, and several police cars escorted the caravan with lights flashing and sirens blaring.

Professor Dumbell personally accompanied the shipment on its flight to the United States. When he arrived at the Atlanta airport, a small reception committee from the CDC was waiting to greet him. Having worked on smallpox all his professional life, Dumbell had mixed feelings about delivering the virus to its new guardians. "It was like saying farewell to an old friend," he observed later. In contrast to Heathrow, the security arrangements in Atlanta were modest: The CDC scientists simply placed the two Styrofoam containers in the trunk of a car and drove away.

In 1983, China reported to the WHO that it had destroyed its stocks of variola virus, leaving South Africa as the only declared possessor outside the United States and the Soviet Union. The National Institute for Virology in Sandringham, near Johannesburg, had a collection of 177 vials containing local isolates of variola virus. Although the virus had not been studied in years, the South African Ministry of Heath had retained the stocks because of concerns that without them, the political isolation of the apartheid regime would make it difficult to diagnose smallpox if it ever reappeared. After Keith Dumbell moved to South Africa and made his library of cloned smallpox DNA available to the National Institute for Virology, however, the South African government finally dropped its objections to destroying the virus stocks. In a special ceremony at Sandringham on December 9, 1983, the strain collection was placed in an autoclave and the South African Minister of Health and Welfare, C. V. van der Merwe, pressed a button to start the destruction process. Exposure to pressurized steam for twenty minutes was sufficient to inactivate the virus.

By early 1984, the WHO had authorized only two laboratories in the world to retain samples of variola virus: the CDC in Atlanta and the Research Institute for Viral Preparations in Moscow. Because these two labs had done all the diagnostic work for the WHO during the smallpox eradication campaign, they had accumulated the largest strain collections. The CDC had a total of 451 viral isolates, including those transferred from Great Britain, Japan, the Netherlands, the U.S. Army, and the American Type Culture Collection (ATCC), a nonprofit organization that maintains cultures of microorganisms for biomedical research. (ATCC had agreed to transfer its stocks to the CDC in February 1979 on the condition that the company be consulted before a decision was made to destroy them.) Of the CDC specimens, twenty-two were original scabs from smallpox patients and the rest were frozen or freeze-dried cultures. The Moscow repository contained single or multiple samples of 120 isolates collected since 1958, including strains from eighteen countries in Africa, Asia, and South America, as well as the Soviet Union. Seventeen specimens were scabs from smallpox patients, ninety were frozen cultures, and twenty-four were in freeze-dried form.

In 1984, the WHO asked all member countries to certify that no stocks of variola virus existed outside the two official repositories. Over the next several months, the organization received numerous letters confirming destruction or transfer. The accuracy of these reports depended, of course, on the good faith of the respondents. Because it was logistically and politically impossible for the WHO to inspect every laboratory freezer in the world and to analyze the contents of each vial, the organization had no choice but to trust countries and research centers to search their culture collections thoroughly and to destroy the specimens containing variola virus. In view of the difficulty of accounting for all samples of the virus, the Smallpox Eradication Unit had chosen to define the endpoint of the smallpox eradication campaign as the absence of human cases of the disease, and not the worldwide elimination of the virus itself.

Nevertheless, the possibility that lost or forgotten samples of variola virus might still exist was a matter of concern. During the eradication campaign, virologists in several countries had isolated the virus from clinical specimens, grown it on egg membrane, and stored some of the cultured material in a deep-freezer so they could compare new isolates with earlier ones. Most virologists were poor librarians, however, and they rarely catalogued or cleaned out their laboratory freezers on a regular basis. Thus, over the years, specimens of variola virus tended to get lost among the mass of ice-encrusted tubes and bottles. Labels fell off vials and scientists died or moved to other institutions, leaving their undocumented culture collections behind.

Given this situation, it was not particularly surprising that a few laboratories that had supposedly destroyed their stocks of variola virus later stumbled across some forgotten samples. On March 23, 1979, for example, staff virologists at the California State Department of Health Services discovered twelve ampules of variola virus whose existence was unrecorded, as the lab had supposedly destroyed all its stocks in 1976. The ampules were autoclaved the same day. Laboratories in Britain and Tanzania also turned up undocumented samples, and a Czech research institute in Prague reportedly ignored the WHO order to eliminate its stocks of variola virus "out of laziness, not out of any evil intentions," and only got around to destroying them several years later. Some experts worried that in the People's Republic of China, samples of variola virus might have been misplaced during the Cultural Revolution of the 1960s, when many scientists and technicians had been imprisoned or killed.

After the 1980 World Health Assembly voted to end the routine vaccination of civilians against smallpox, most countries discontinued their vaccination programs by 1982 and the last one—France—did so in 1984. Nevertheless, certain groups continued to receive the vaccine, including scientists in more than fifty laboratories that conducted research with animal poxviruses or vaccinia. In addition, fears that smallpox might be used as a biological weapon led several countries—including Australia, Canada, Israel, West Germany, several other European NATO allies, the United States, and the Soviet Union—to keep vaccinating their military personnel. U.S. recruits and national guardsmen were vaccinated against smallpox when they entered the service and revaccinated at five-year intervals; about a million American soldiers received the smallpox vaccine each year.

According to Soviet Deputy Minister of Health Pyotr N. Burgasov, the Soviet Union stopped vaccinating its troops against smallpox in 1979 but started again in 1984. Although Burgasov gave no reason for this decision, it was presumably related to the fact that the U.S. military was continuing to vaccinate. In 1984, the *New York Times* editorial page called on both superpowers to halt military smallpox vaccination, arguing that it was "difficult to imagine any purpose for which a country would be willing to incur the odium of reintroducing the disease." Ending the vaccination of soldiers, the *Times* argued, would build mutual confidence and represent a "final step in ending the fear of smallpox."

By 1986, ten countries—including Belgium, Great Britain, the Netherlands, Norway, Finland, and Sweden—had stopped vaccinating their military recruits against smallpox. In March 1990, the United States did likewise, not because of a change in the perceived biological warfare threat, but because of concerns about the stocks of vaccinia immune globulin (VIG), the antiserum used to treat the occasional severe complications caused by smallpox vaccine. Canada and Russia also halted the vaccination of regular troops against smallpox in the early 1990s, leaving Israel as the only country to continue doing so, at least openly. Reportedly, the Pentagon continued in secret to vaccinate small groups of special forces involved in covert operations against rogue states and terrorist organizations.

Because the contagiousness of smallpox made it difficult to target or control, it appeared to be an unlikely biological weapon. In 1992, however, a high-ranking Soviet official defected to the United States and gave the U.S. intelligence community some chilling news. He reported that in parallel with the global WHO campaign to eradicate smallpox—an effort in which Soviet virologists, epidemiologists, and vaccine manufacturers had played a leading role—the Soviet military had cynically pursued a top-secret program to transform the virus into a doomsday weapon.

EIGHT

THE SOVIET BETRAYAL

The British use of smallpox as a weapon during the eighteenth century set the stage for renewed military interest in the virus in the mid-twentieth century. During World War II, British, Canadian, and American scientists studied smallpox as a potential biowarfare agent, although they eventually abandoned this line of research because the contagious nature of the disease made it difficult to control. Japanese army doctors, however, exposed Chinese prisoners of war to aerosolized variola virus at a secret germ research center in occupied Manchuria known as Unit 731. According to a Chinese account, "Those experimented upon developed pustules all over their bodies, and pleaded 'Save me!' until they were hoarse. They were left to die untreated and their bodies were burned." After World War II, the U.S. Army's biological warfare center at Fort Detrick, Maryland, did additional research on smallpox before President Richard M. Nixon officially halted the U.S. offensive program in November 1969. Even so, variola virus was never particularly high on the list of agents that the army considered worthy of investment because vaccination provided an effective defense against military use.

The Soviet Union's program to develop smallpox as a biological weapon far outstripped the efforts of other countries in both scale and sophistication. Beginning in 1928 and continuing throughout World War II, Soviet military scientists did research on a variety of disease agents at a secret biowarfare facility on the Solovetsky Islands in the White Sea, but weapons based on smallpox were not developed until after the war. In 1947, the Soviet Ministry of Defense established the Center of Virology near Zagorsk, a picturesque city of monasteries and onion-dome churches some forty-five miles northeast of Moscow. Inside a walled army compound on the outskirts of the city, military scientists attempted to turn several viruses into biological weapons. Variola major was considered the agent of choice for inflicting mass casualties because it was highly infectious through the air, was rugged enough to survive explosive delivery, and caused a debilitating and demoralizing disease with high mortality.

Prior to mass-producing the smallpox biological weapon, it was necessary to select the strain of variola major to be cultivated. The Soviet intelligence services acquired numerous strains of the virus through covert operations, such as stealing seed cultures from foreign laboratories, buying them from corrupt scientists, or ordering them through undercover agents posing as legitimate researchers. In addition, Soviet physicians who traveled to other countries to assist in combating natural outbreaks of smallpox brought back samples of the virus for the culture collection of the Ministry of Health, to which the Ministry of Defense was granted full access. At the Center of Virology in Zagorsk, Colonels Igor Nikonov and V. V. Zezerov evaluated dozens of strains of variola major for the best combination of militarily relevant characteristics, including low infectious dose and stability when dispersed as an invisible aerosol cloud.

In 1959, an Indian visitor to Moscow imported a particularly virulent and contagious form of smallpox, causing an outbreak that lasted forty-four days and sickened forty-six people. Were it not for the prompt vaccination of 6.6 million residents of Moscow within a three-week period, the epidemic would have spread widely. This incident inspired the Soviet government to send a special medical team to India in 1967 to help combat smallpox in that country, an action that was widely heralded in the official state media. Unbeknownst to the world, however, the Soviet physicians were secretly accompanied by KGB agents who obtained a sample of the Indian strain of variola major. Designated "India-1967," it appeared to be particularly well suited for weapons production: It killed more than 30 percent of those infected, retained its virulence when stored for long periods, and was extremely stable in aerosol form. Because of these advantages, the Soviet army adopted the India-1967 strain (later code-named "India-1") for all subsequent biological weapons production. Although comparison testing with other strains of variola major continued for several years, no other strain was found to be superior for military purposes.

Whereas bacteria can be grown in a nutrient-rich broth, viruses replicate only in living cells and hence are more difficult to manufacture in large quantities. Soviet army scientists at Zagorsk, led by Colonels Nikonov and Zezerov, perfected the standard method for producing variola virus in fertilized chicken eggs. This technique involved cutting a small window in the eggshell and using an extremely fine needle to inject a minute amount of the virus into the developing chick embryo. The infected egg was then sealed with paraffin and warmed in an incubator, allowing the virus to multiply prolifically within the chorioallantoic membrane of the embryo. After three or four days, the egg membranes were harvested and homogenated, yielding a liquid suspension so enriched in variola virus that it usually did not require further concentration.

As a rule, Soviet bioweaponeers preferred to manufacture bacterial or viral agents as a dry powder rather than as a wet slurry, because the dried agent had a significantly longer shelf life and could be disseminated more efficiently as an aerosol. This principle did not apply to the smallpox weapon, however, because the liquid smallpox formulation retained its viability for months when deep-frozen and was extremely stable in aerosol form. Moreover, dried smallpox virus posed an extreme hazard of infection to workers during the manufacturing process. For these reasons, the Soviets always produced the virus in liquid form.

Technicians at the Center of Virology mass-produced the India-1967 strain by inoculating the virus into thousands of fertilized chicken eggs, which were warmed in room-size incubators. Vast numbers of eggs from state-run collective farms were secretly diverted to the production line at Zagorsk; because the Soviet economy was centrally planned, the eggs could be removed from the civilian market without public knowledge. To permit the manufacture of even larger quantities of variola virus in wartime, a backup production line using fertilized eggs was established at the Pokrov Biologics Plant, a Ministry of Agriculture facility about 200 miles southeast of Moscow.

Once manufactured in eggs, the concentrated viral suspension was converted into a finished product by mixing it with a complex formula of chemical additives, including a stabilizer to prolong the viability of the virus in storage and an inert filling agent to facilitate its dispersal as a fine-particle aerosol. Another additive served to lower the freezing point of the viral suspension so that it could be stockpiled at sub-freezing temperatures (down to −40 degrees F.), significantly extending its shelf life. The total amount of time required to manufacture weapons-grade smallpox from the initial egg inoculation to the finished product was on the order of one week. To disseminate the smallpox agent, specialized bomblets were developed for delivery in aerial bombs and ballistic missile warheads. Field testing of smallpox biological weapons took place at an open-air test site on Vozrozhdenie Island in the Aral Sea, although the danger of epidemics caused the Soviets to halt such experimentation in the early 1970s.

During the late 1960s and early 1970s, the Strategic Rocket Forces deployed single warheads containing bomblets filled with variola virus on SS-11 intercontinental ballistic missiles, which were based in silos in the northern latitudes of the Soviet Union. The cold temperatures in the far north kept the stored smallpox agent viable for long periods. Soviet engineers later developed special refrigerated warheads to enable the biological payload to survive the intense heat of reentry through the atmosphere. Each warhead contained an internal cooling system that, when combined with thermal shielding and the spinning of the reentry vehicle, kept the temperature inside the warhead well below the boiling point. Toward the end of its ballistic trajectory, the smallpox warhead would deploy a parachute that slowed its velocity. At the appropriate altitude, the warhead would break open, releasing its payload of bomblets to disperse a deadly cloud of aerosolized virus over the target area.

Because biological warfare agents, including smallpox, had an incubation period of days to weeks, they were not suitable for tactical use on the battlefield. Instead, Soviet military planners reserved such weapons for operational and strategic missions. Operational biological weapons were designed for attacking midrange targets, including NATO military bases in Western Europe, and were based on agents that caused an incapacitating but usually nonfatal illness, such as Venezuelan equine encephalitis (VEE) virus and the bacterial agents glanders and tularemia. Strategic biological weapons, in contrast, were intended to strike long-range targets in the United States and, after 1968, the People's Republic of China. These weapons were based primarily on contagious and often lethal microbes, such as plague bacteria and smallpox virus. (A few agents, such as anthrax bacteria and Marburg hemorrhagic fever virus, were considered suitable for both operational and strategic use.) Soviet military doctrine for strategic biological warfare called for delivering massive quantities of contagious agents against urban targets to cause panic and social disruption, overwhelm the enemy's medical system, and spawn widespread epidemics that would be impossible to control. This philosophy of overkill was reflected in the size of Soviet production facilities, which had the capacity to manufacture tens to hundreds of tons per year of smallpox, plague, and anthrax microbes.

Smallpox biological weapons were intended for use against U.S. cities in a war of total mutual annihilation, with the aim of killing the survivors in the aftermath of a nuclear exchange. Although smallpox normally had a two-week incubation period, individuals exposed to a cloud of aerosolized virus released by a weapon would receive many times the dose encountered under natural conditions. It was believed that this concentrated exposure would accelerate the infection process. As a result, the incubation period might be as short as one to five days, depending on the dose and the immune status of the host. It was also likely that people infected with smallpox by secondary spread would rapidly develop acute illness because radioactive fallout from nuclear explosions would suppress the immune system and reduce the body's ability to fight infection.

In 1972, the Soviet Union signed the Biological Weapons Convention (BWC), an international treaty banning the development, production, and stockpiling of weapons based on disease agents and natural poisons. Largely because of Soviet objections to on-site inspections, this treaty included no measures for checking compliance and hence was little more than a gentleman's agreement. After Moscow signed the treaty but before it had been ratified by the Soviet parliament, an internal debate took place within the Politburo over whether or not to abandon the offensive biowarfare program. Leading biological scientists, such as Dr. Yuri Ovchinnikov, the vice president of the Soviet Academy of Sciences, suggested that it would be possible to circumvent the BWC by using microbes to produce biologically active substances that could replace classical chemical weapons; such production could be concealed within the biotechnology and pharmaceutical industries. Ovchinnikov also envisioned the use of genetic engineering to develop a new generation of microbial agents with an enhanced capacity to express toxins and other biologically active substances, thereby improving their military effectiveness. In view of these promising technological developments, the Politburo decided not only to maintain the offensive biowarfare program but to intensify it.

Soviet leaders rationalized the development of a new generation of biological weapons as a necessary counterweight to the perceived U.S.-NATO military threat. Despite President Nixon's official renunciation of biological warfare in 1969, the Kremlin believed that Washington was secretly continuing to develop biological weapons under the cover of defensive research. Although this assessment was wrong, it is now known that Soviet leaders had some basis for making it. During the 1970s, U.S. military intelligence used double agents, including Sergeant Joseph Cassidy and Dmitry Polyakov, to feed false information to Moscow that the United States was continuing in secret to develop new generations of chemical and biological weapons. The purpose of this disinformation campaign was to encourage the Soviets to squander resources on toxic weapons of limited military utility. Ironically, Soviet military scientists, spurred on by the perceived need to catch up with the United States, made several unexpected breakthroughs. They successfully developed a new family of supertoxic nerve gases known as *novichok* (Russian for "newcomer") and a variety of "improved" germ warfare agents. Thus, a dangerous miscalculation by U.S. military intelligence was partly to blame for the terrifying achievements of the Soviet biological warfare program.

Because the Biological Weapons Convention banned the development, production, and stockpiling of pathogens and toxins for warfare purposes, the Soviet offensive program would henceforth have to be shrouded in the deepest secrecy. To this end, in 1973 the Politburo established a new Special Main Directorate under the Council of Ministers of the Soviet Union known as "Biopreparat." To the outside world, Biopreparat was a state-owned pharmaceutical complex that developed drugs and vaccines for the civilian market. In reality, it was an elaborate front for a military-funded program code-named *Fermenty* (the Russian word for "enzymes"), which aimed to develop a new generation of superlethal biological weapons. Legitimate civilian work provided a cover for the offensive biowarfare program and never accounted for more than 15 percent of the activity of the Biopreparat research institutes and production plants. Because *Fermenty* was a clear violation of international law, it remained one of the Soviet Union's most closely guarded secrets, and only a small circle of senior Kremlin officials understood its full scope.

Overseeing the *Fermenty* program was a government coordinating body known as the Inter-Agency Scientific and Technical Council on Molecular Biology and Genetics. In 1973, Alexei Kosygin and Leonid Brezhnev appointed as the first chairman of the council Dr. Viktor M. Zhdanov, a well-known virologist and senior member of the Soviet Academy of Sciences. Ironically, this Zhdanov was the same man who, as deputy minister of health, had first proposed the global eradication of smallpox at the 1958 World Health Assembly in Minneapolis. After chairing the inter-agency council for two years, Zhdanov was forced to resign in 1975; the reason for his departure was poor managerial ability, rather than a lack of enthusiasm for the work. He spent the remainder of his career doing peaceful research on hepatitis, influenza, and AIDS.

During the 1970s, smallpox was considered such a key element of the Soviet biological arsenal that the military command issued an order to maintain a stockpile of up to twenty metric tons of variola virus suspension at the Center of Virology in Zagorsk. The capacity of the egg production method was limited, however, by its extensive demands for material and manpower. Although the technique had gradually been made more efficient by developing automated inoculation and incubation equipment, certain steps, such as checking the suitability of embryos for inoculation, could not be mechanized with existing technology. Accordingly, Colonel Yevgeny Lukin and General S. I. Prigoda at Zagorsk began to develop novel production techniques involving the use of stainless steel bioreactors, such as those employed for the manufacture of vaccines. Seed cultures of variola virus were inoculated into one-liter roller bottles containing a single layer of tissue cells; when the virus had multiplied inside the cells, the cells were transferred into small-scale bioreactors (100 to 250 liters), in which they were suspended in a liquid nutrient broth.

Although the viral product obtained by the cell-culture method was less concentrated than that produced in fertilized eggs, it was usually adequate for manufacturing the final formulation used in smallpox biological weapons. Numerous technical difficulties, however, prevented the bioreactors from fully replacing the egg method, so that both techniques were used simultaneously. The viral suspension was stored at the army facility in Zagorsk in two types of refrigeration tanks, one containing 50 kilograms of agent and the other 250 kilograms. Since variola virus remained potent in storage for between six months to a year, annual production quotas were needed to replenish the existing stocks as they decayed.

In wartime, the smallpox agent would have been loaded into various types of strategic delivery systems, including aerial bombs delivered by long-range bombers and warheads for intercontinental ballistic missiles targeted on U.S. cities. The twenty tons of smallpox virus formulation stockpiled at Zagorsk were sufficient to cover approximately 4,000 square miles of territory. Assuming that the target area was densely populated, the number of people exposed would have been in the millions. Given the extremely low infectious dose of variola virus, the short incubation period after exposure to a concentrated aerosol, the contagiousness of the disease, and the fact that the corpses of the dead would have remained infectious for days, a Soviet attack with smallpox weapons against urban targets would have dealt a devastating blow to the United States, perhaps destroying it as a functioning society.

In 1980, when the WHO formally certified the global eradication of smallpox and decided to halt the vaccination of civilians, the Kremlin cynically viewed this triumph of international public health as a military opportunity. From now on, the world's population would become increasingly susceptible to smallpox infection, transforming the virus into a true weapon of mass destruction. At this juncture, the Soviet leadership ordered the development of an "improved" version of the smallpox weapon that had been in the country's arsenal for decades. The Communist Party of the Soviet Union and the Soviet government issued a five-year plan for Biopreparat for the period 1981–85 that called for the intensive study of certain viruses, including variola major, to develop more lethal strains.

Despite the risk that military use of a contagious agent like smallpox might eventually boomerang against the Soviet Union, the Kremlin made no attempt to vaccinate its own civilian population after 1980. Beyond the need to preserve the secrecy of the offensive biowarfare program, smallpox was intended for use in a war of mutual destruction that few people were likely to survive. Indeed, in 1984–85, Soviet scientists carried out a systematic search for new stabilizing chemicals that would shield variola virus from ionizing radiation. The technical specification stated that these additives were needed for the effective utilization of the smallpox weapon in the highly radioactive environment that would exist in the aftermath of a nuclear exchange.

Kanatjan Alibekov played a key role in the history of the Soviet biological warfare program and, in particular, the development of the "improved" smallpox weapon. Born in 1950 in Kauchuk, a small town in the Soviet republic of Kazakhstan, he was raised in the capital of Alma-Ata. His father was an ethnic Kazakh who worked as a career militia officer, and Kanatjan's straight black hair, broad face, light brown skin, and faintly Oriental features reflected his Central Asian ancestry. His first language was Kazakh, although he learned Russian in school. At age seventeen, aspiring to become a physician, he entered a six-year program at the medical institute in Alma-Ata. After four years, he decided to specialize in military medicine and enlisted in the Soviet army. He then became a cadet at the medical institute in Tomsk, Siberia, where he specialized in infectious diseases and epidemiology.

Shortly before his graduation in June 1975 at the age of twenty-four, Alibekov was interviewed repeatedly by scientists and federal security officers, who told him that he was being considered for an important scientific position requiring a high-level clearance. Although he had no idea what type of work he would be asked to perform, he was eager to contribute to the nation's defense. The assignment turned out to be with Biopreparat. In August 1975, Alibekov received his first assignment at a standby production plant near the Russian city of Omutninsk, where he spent seven months learning the basics of industrial microbiology and biological weapons development. At first, he was tormented by doubts. Employing disease as a method of warfare violated the Hippocratic oath that he had taken as a physician: "First, do no harm." But Alibekov was as patriotic as the next man, and he came to accept the propaganda line that the United States and its NATO allies were secretly developing biological weapons and that the Soviet Union had to do the same to defend itself.

Particularly for a Soviet citizen of non-Russian ethnicity, Alibekov enjoyed a meteoric rise through the ranks of Biopreparat. After his initial training at Omutninsk, he worked at another bioweapons plant in Siberia from 1976 until 1980. He then returned to Omutninsk as chief of the technical department, where he developed a weapon based on the bacterial disease tularemia. In August 1983, he became director of a vast research and production facility in Stepnogorsk, on the windswept Kazakhstani steppes, where he supervised the development of a new and more deadly strain of anthrax. For his work at Stepnogorsk, he was promoted to colonel two years ahead of schedule.

Finally in September 1987, at age thirty-seven, Alibekov was called to Moscow to become chief scientist and first deputy director of the entire Biopreparat complex. By this time, the Soviet germ warfare program was formidable in both scale and scope. The Kremlin knew that it could not match the U.S. military buildup of the 1980s, including the Reagan administration's ambitious plans for a space-based "star wars" defense against ballistic missiles. Although biological warfare was not a direct response to the U.S. buildup, it was seen as an area of compensatory Soviet advantage.

In addition to four military microbiological institutes operated by the Soviet Ministry of Defense, roughly forty Biopreparat facilities employed more than thirty thousand men and women, most of them engaged in research and development on a variety of biological weapons. The Soviet germ arsenal included multiton stockpiles of the causative agents of smallpox, anthrax, and plague, and seven mothballed production facilities were kept in reserve for the manufacture of more deadly microbes in wartime.

Three months after his arrival at Biopreparat headquarters in Moscow, Alibekov was assigned by his boss, General Yuri T. Kalinin, to supervise the development and production of an "improved" smallpox weapon. This project was included in the 1986–90 five-year plan for Biopreparat that Communist Party General Secretary Mikhail Gorbachev had signed in February 1986, several months after taking office. Funding for the five-year plan would total more than $1 billion by the end of the decade, enabling the Soviet Union to catch up with, and even surpass, the West in certain areas of biotechnology. Among other objectives, the plan called for the development of a pilot plant for manufacturing the India-1967 strain of variola major, with the aim of expanding production from small-scale to large-scale bioreactors.

Most of the development work on the new smallpox weapon took place at the All-Union Scientific Research Institute of Molecular Biology in Koltsovo, a small town in western Siberia about twenty miles southeast of Novosibirsk. Known to initiates as "Vector," the top-secret laboratory was funded by the Soviet state as part of the Biopreparat system and specialized in developing viral agents for biological warfare. Although the first buildings at Vector were built by prison labor in 1975, it was not until 1982 that the construction of special maximum-containment laboratories permitted work with extremely dangerous viruses. The center also had a culture collection of about ten thousand viral strains, including deadly hemorrhagic fever viruses such as Ebola, Marburg, Lassa, and Machupo.

Some 4,500 people, including about 250 Ph.D.-level scientists, worked at Vector in the late 1980s. The center comprised nearly fifty research, production, and administration buildings in a bland international style, some of them several stories high. Ringing the complex was a reinforced concrete wall, with security checkpoints at the entrances, surrounded by a buffer zone of open fields sparsely planted with silver pines. According to a description in the Russian press, "Vector . . . has the ineradicable stamp of a military institution—there are as many comforts or semblances of design . . . as in a tank turret. Everything here is square, rectangular, of good quality, and exclusively functional."

Secret bioweapons work at Vector was protected by two separate cover stories, one for public consumption and the other for internal use. Outsiders were led to believe that the center had been established for the creation and production of biological pesticides for use in agriculture. Most of the scientists working at the center knew that that their main client was the Ministry of Defense, but they were told that the purpose of their work was to develop effective means of protecting Soviet troops against biological weapons of mass destruction. Only a few senior scientists knew the truth: that Vector had been created for the design and production of offensive viral weapons.

The center's director was Dr. Lev Stepanovich Sandakhchiev, a biochemist of Armenian ethnicity. A small man with an olive complexion, graying hair, and teeth bearing numerous gold fillings, he appeared cordial and engaging to outsiders, but in fact he was a chameleon whose behavior changed with his surroundings. Trained in Moscow as an organic chemist, Sandakhchiev had studied for his doctorate in the biochemistry department of the Novosibirsk Institute of Organic Chemistry. After coming to Vector in 1975 at the age of thirty-seven, he had become interested in the emerging field of molecular virology, including the genetic engineering of viruses. Despite a lack of major scientific achievements, his success as a manager and his political connections had enabled him to rise to the position of active member of the Soviet Academy of Sciences, the highest rank bestowed on a Soviet scientist.

Under Sandakhchiev's energetic leadership, Vector thrived and expanded. Although he personally considered biological weapons work uninteresting, he knew that his future as a manager depended on how well the center carried out the tasks assigned to it by the Soviet state. Accordingly, he personally supervised all the scientific and technical developments in the bioweapons field and made sure that the orders from Biopreparat director Kalinin were fulfilled on schedule. Under the 1986–90 five-year plan, several new facilities were built at Vector, including a maximum-containment laboratory suite for work with variola and other deadly viruses, explosive chambers for testing biological munitions, and animal houses.

Most of the development work on smallpox took place in two large, identical-looking brick-and-concrete structures known as Buildings 6 and 6A, which were completed in 1986. Each building was four stories high, with courses of exterior windows corresponding to the corridors ringing the maximum-containment laboratories. The two buildings were surrounded by a security fence and guarded around the clock by soldiers from a special military division.

Building 6 contained 1,440 square meters of maximum-containment laboratory space known as "Zone 3," where Soviet scientists wore plastic space suits equipped with self-contained air supplies to isolate them from the deadly viruses they handled. The interior design of the laboratory was heavy and functional, reminiscent of a World War II battleship, with steel walls and submarine doors. Special ventilation systems maintained negative air pressure, so that a breach of containment would not allow dangerous pathogens to escape into the environment, and three redundant power supplies ensured that the containment system would function under all circumstances. The building also had a large aerosol test chamber and an animal room housing rodents, rabbits, monkeys, and baboons.

Building 6A had a labyrinthine series of rooms containing aerosol test chambers, the largest of which was a cube ten meters on each side. A centralized aerosol generator with a computerized control panel sent a stream of virus-laced air through stainless-steel pipes to the various test chambers. At the flick of a switch, the viral aerosol could be pumped through a ceiling vent into one of the chambers, exposing the caged experimental animals inside—rodents, sheep, or primates. Some of the chambers were made of reinforced steel to permit the testing of explosive bomblets.

Building 6A also had numerous small test chambers, called "crocodiles," each consisting of a sealed glove box mounted on a pedestal so that technicians could observe it from all sides. Within the glass-enclosed chamber, a cloud of aerosolized virus particles circulated in a path resembling a flattened doughnut, enabling scientists to study the decay of the viral agent over time as they modified the conditions of heat, humidity, and light. It was also possible to insert the head of a monkey or a rodent through an opening in the side of the chamber to test the biological effects of the virus.

A key task specified for Vector in the 1986–90 five-year plan was to develop and refine a process for producing variola virus in large bioreactors. Not only was growing the virus in cell culture more efficient than cultivating it in chicken eggs, but it would also be easier to conceal under the cover of a legitimate commercial activity, such as vaccine production. Because Sandakhchiev knew little about the large-scale manufacture of human viruses, he requested the transfer to Vector of two military virologists, Colonel Yevgeny Lukin from the Center of Virology in Zagorsk and Lieutenant Colonel Yevgeny Stavsky from another Ministry of Defense facility in Sverdlovsk. Colonel Lukin, a man in his early fifties who carried himself with military bearing, had a doctorate in technical sciences and specialized in industrial microbiology; at Zagorsk, he had worked for several years on smallpox production methods. Colonel Lukin and his deputy, Dr. Vladimir Shishkov, were assigned responsibility for the production and testing of viral agents at Vector. Lieutenant Colonel Stavsky, holder of a Ph.D. in microbiology, headed the Smallpox Weapon Development Department in Building 6.

Under the direction of Lukin and Stavsky, a team of fifteen Vector scientists and technicians worked from 1987 to 1990 to scale up the process developed at Zagorsk for growing variola virus in cultured cells. They employed a pilot-scale bioreactor with a capacity of 630 liters, which was installed inside a large walk-in incubator in Building 6. The bioreactor was a cylindrical stainless steel tank about five feet high, with an external water blanket to regulate the temperature of the nutrient broth within and a rotating agitator blade at the bottom of the tank to keep the animal cells continually suspended in the liquid. Pipes fed in fresh nutrient solution and removed wastes, and a porthole in the top of the tank allowed the scientists to peer inside.

In developing the production process, the smallpox team at Vector made use of a highly classified, multivolume manual compiled by the Center of Virology in Zagorsk. This manual provided detailed recipes for the complex nutrient media and culture conditions needed to grow the virus, and the formula of chemical additives used to extend its shelf life in storage. Access to the smallpox manual was tightly controlled by the Soviet Ministry of Defense, which made sections available to Vector scientists only at Colonel Lukin's request and on a strict "need to know" basis.

Alibekov's greatest fear during the smallpox research program at Vector was that small amounts of variola virus might escape from Building 6 and cause an epidemic in the surrounding area. Now that smallpox had been eradicated worldwide, an outbreak of the disease would be impossible to explain as a natural phenomenon and would thus arouse intense suspicion. For this reason, Vector went to extraordinary lengths to ensure effective biocontainment. Everyone who worked with smallpox received a vaccination on the buttocks to avoid creating a fresh scar on the upper arm, and blood antibody levels were monitored frequently to ensure a high level of immunity. In addition, team members were isolated from the outside world for three or four months at a time. They ate and slept in special dormitories inside Building 6, drew hazardous-duty pay, and received time off to visit their families. Before team members could leave the building, they had to undergo a thorough medical examination and remain in quarantine for two weeks, the incubation period of smallpox. All of these procedures were conducted in such a way as not to attract outside attention.

The Vector team finally succeeded in growing variola virus in the bioreactor in cultured monkey kidney cells, although the yields were significantly lower than in chicken eggs. In December 1990, the new smallpox formulation was tested successfully in an aerosol test chamber in Building 6. Because open-air experimentation was impossible, the testing method consisted of exploding miniaturized models of bomblets filled with smallpox agent inside the steel-walled chamber; using mathematical models, scientists then extrapolated the results to predict the effects of a full-scale weapon. In recognition of their achievements, the developers and testers were nominated for the State Prize (although the breakup of the Soviet Union in December 1991 prevented them from receiving it).

During the next phase of the project, Lukin and Stavsky planned to establish a mobilization capability to manufacture multiton quantities of variola virus suspension in a series of bioreactors with capacities ranging up to 2,500 liters, which had been installed in a newly built production facility known as Building 15. It was anticipated that the new manufacturing line, to be activated only in wartime, would have a capacity of between eighty and one hundred tons of viral suspension per year.

According to the Soviet master plan, during the period of tension and crisis preceding World War III, the military command would issue a special order to mass-produce the smallpox agent and load it into munitions. The viral suspension manufactured at Vector would then be transported by rail to another mobilization facility in the nearby town of Berdsk, where a filling machine similar to one in a soft-drink bottling plant would inject the liquid agent into special bomblets. Resembling an oblong grapefruit, each bomblet consisted of two agent-filled aluminum hemispheres surrounding a small, egg-shaped explosive charge. In wartime, bomblets filled with smallpox virus suspension would be loaded into cluster bombs for delivery by TU-95 long-range strategic bombers; a cluster bomb held roughly 110 bomblets.

The other delivery systems for smallpox were intercontinental ballistic missiles (ICBMs). Until the late 1980s, single-warhead ICBMs with biological payloads were deployed near the Arctic Circle. In December 1988, however, the Soviet Ministry of Defense decided to use the giant SS-18 intercontinental ballistic missile, previously armed only with nuclear weapons, to carry biological agents. An SS-18 could deliver ten independently targetable warheads over a range of six thousand miles. Each smallpox warhead contained about 150 bomblets; at a predetermined altitude above the surface, the casing would break open and disperse the bomblets in a large "footprint" over the target. A fixed propeller would cause each bomblet to spin as it fell to earth; at about twenty-five to one hundred meters above the ground, the explosive charge would detonate, rupturing the thin aluminum walls and breaking up the liquid suspension into tiny droplets. The exploding bomblets would create multiple point sources over the target that would merge into a vast infectious cloud. One SS-18 missile could deliver a total of 375 kilograms of viral suspension, or enough to cover a target area of 75 to 150 square kilometers. Several other delivery systems for smallpox were studied but not implemented, including high-speed cruise missiles that disseminated the agent with various spraying techniques.

At the same time that the Lukin-Stavsky team was developing an improved method for mass-producing variola virus, scientists in the smallpox research division in Building 6 were exploring the application of genetic engineering techniques to create a new generation of smallpox weapons. Invented in the early 1970s, genetic engineering involves the use of bacterial enzymes to cut and splice molecules of DNA, with the aim of transplanting genes from one species to another. The potential misuse of this powerful technology for warfare and terrorism has been termed "black biology."

In 1973, the Soviet government issued a decree to accelerate the country's research effort in molecular biology and genetic engineering. Although the new initiative was ostensibly intended for peaceful purposes, a top-secret annex stated that the Soviet Union should also use the new technologies to develop advanced biological weapons, including novel variants of agents already in the Soviet arsenal. Pursuant to this decree and subsequent government decisions, the Soviet Academy of Sciences began work on new methodologies for genetic engineering. Vector then applied these techniques with the goal of creating genetically altered strains of smallpox for use in biological weapons.

To minimize the risk that the offensive research program at Vector would be detected, Alibekov decreed that all efforts to develop next-generation smallpox weapons should be disguised as legitimate research by employing vaccinia or animal poxviruses as models. Research with vaccinia offered an ideal cover story because the virus was then being studied intensively around the world as a vehicle for genetically engineered vaccines, in which genes from other pathogens were inserted into vaccinia as a means of inducing immunity. Once the basic development work on a new viral weapon had been carried out with vaccinia or another surrogate virus, just a few weeks would be needed to perform the same manipulations with variola and complete the final testing. To provide a knowledge base for efforts to "improve" the smallpox weapon, Vector scientists were ordered to analyze the structural and functional organization of the variola major genome, meaning the full complement of viral genes. It would then be possible to compare the genetic structure of variola with those of vaccinia, monkeypox, and mousepox to identify the genes responsible for the virulence and infectivity of smallpox in humans.

The program at Vector to develop genetically engineered smallpox weapons was divided into subtasks. One subtask was to develop new strains of variola that would be resistant to existing vaccines. The planned approach was to identify the common proteins in vaccinia and variola that induced immunity, and then to develop engineered strains of variola containing modified viral proteins that the human immune system could not recognize and defend against. A second subtask was to develop more lethal forms of smallpox by inserting foreign genes into the variola DNA, including those coding for small protein toxins. This work was orchestrated by Dr. Stanislav Vasilyenko, head of the department of molecular genetics, and Dr. Vladimir Kravchenko, head of the genetic engineering laboratory. A third subtask was to clone fragments of India-1967 in bacteria so that the viral DNA would be available to Soviet researchers even after the stocks of live variola virus were destroyed, perhaps making it possible to reconstitute the virus in the laboratory.

In February 1989, during one of Alibekov's inspection trips to Vector, Lev Sandakhchiev introduced him to Sergei V. Netesov, an ambitious young scientist in his thirties who had graduated from Novosibirsk State University in 1975. Netesov explained that he was trying to splice genes from one virus into another to create hybrid viruses called "chimeras," after the beast in Greek mythology with the head of an eagle and the body of a lion. In principle, such chimeras would combine the militarily relevant traits of different viruses, such as infectivity, lethality, and robustness. Impressed by Netesov's initiative, Alibekov authorized Sandakhchiev to promote him to the position of deputy scientific director of Vector. He also approved a special grant of 100,000 rubles for a new research program on chimeric viruses, which was code-named *Okhotnik* ("hunter").

As a first step toward the genetic engineering of smallpox, scientists at Vector demonstrated the feasibility of inserting foreign genes into vaccinia without a loss of virulence. Several types of gene transfer experiments were attempted. Dr. Sergei Shchelkunov cloned and inserted into vaccinia a gene for hemorrhagic factor, which interferes with blood clotting, while a team led by Dr. Oleg Serpinsky transferred genes for two morphinelike brain peptides (beta-endorphin and dynorphin), as well as conotoxin, a deadly peptide toxin produced by a marine snail, and fragments of myelin basic protein, which induces an allergic reaction that can cause severe brain damage. In March 1990, Netesov reported that he had inserted into vaccinia a gene from Venezuelan equine encephalitis (VEE) virus, which causes an incapacitating brain disease. Vector scientists also created a chimera of VEE and mousepox virus that was stable and, when used to infect mice, appeared to cause the symptoms of both diseases.

After the gene insertion experiments had been completed successfully with vaccinia, the scientists began similar manipulations with variola major. One goal of the *Okhotnik* program was to develop a smallpox-based biological weapon containing virulence genes from Ebola hemorrhagic fever virus. At least theoretically, such a viral chimera would combine the hardiness and transmissibility of smallpox with the lethality of Ebola, which was between 90 percent and 100 percent fatal, resulting in an "absolute" biological weapon.

Few of the Vector scientists involved in the *Okhotnik* program shared Netesov's enthusiasm for the work. Although Shchelkunov had a personal motivation because he was doing research for his doctoral thesis, the other scientists performed the experiments simply because they were ordered to do so. They did not, however, conduct any large-scale testing with the genetically engineered viruses to determine their stability in aerosol form or their biological effects in nonhuman primates such as monkeys or baboons.

For much of the 1980s, Western intelligence services remained unaware of the full scope and sophistication of the Soviet biological warfare program. In October 1989, however, a senior Biopreparat scientist, Dr. Vladimir Pasechnik, defected to Great Britain while on an official visit to France. During extensive debriefings by British intelligence, he described the

Biopreparat complex and disclosed that his laboratory in Leningrad had developed specific recipes for stabilizing, drying, milling, and disseminating formulations of several microbial pathogens. Pasechnik's revelations stunned the British and American governments. In April 1990, President George Bush and Prime Minister Margaret Thatcher approved a démarche, or formal diplomatic protest, which the U.S. and British ambassadors in Moscow delivered jointly to the Soviet Foreign Ministry. The two ambassadors said that their governments had obtained "new information" that the Soviet Union was conducting an offensive bioweapons program in violation of international law. Taken by surprise, Soviet premier Mikhail Gorbachev denied the allegations out of hand.

Pasechnik's revelations were classified at a high level and made available to only a small number of senior British and American officials. Behind the scenes, however, persistent diplomatic pressure on Moscow finally yielded results. During a U.S.-Soviet summit in Washington in the summer of 1990, Bush and Gorbachev held a private meeting at which the U.S. president raised his concerns about the Soviet biowarfare program. While not admitting that Moscow was violating the Biological Weapons Convention, Gorbachev offered to host a series of unpublicized visits by U.S. and British technical experts to build confidence that the Soviet Union was complying with its treaty obligations. The United States asked to visit four of the Biopreparat institutes. As a face-saving gesture, Gorbachev won the right for Soviet experts to conduct reciprocal inspections of biological facilities in Great Britain and the United States that Soviet intelligence agencies considered to be of possible compliance concern.

Alibekov was responsible for preparing the four Biopreparat facilities selected for inspection by the U.S.-British team, including Vector, by hiding equipment and activities that might be indicative of offensive research and development. For several tense, frantic weeks, he and the various program managers devoted long hours to developing plausible cover stories and concealment strategies. At the same time, the U.S.-British initiative had planted the first seeds of doubt in the back of Alibekov's mind. Why, he asked himself, would the United States and Britain pressure the Soviet Union to halt its offensive biowarfare program if they were engaged in similar activities?

During the inspection of Vector on January 14, 1991, the team of U.S. and British biowarfare experts toured Building 6A. Most of the workforce had been sent home for the day and the laboratories were nearly deserted. Lab benches and equipment had been thoroughly disinfected with bleach and formaldehyde to eliminate any incriminating traces, and an acrid odor hung in the air. British inspector David Kelly saw a technician in one of the labs and took him aside. Speaking through an interpreter, he asked what viral agent they had been studying. "Variola major," the young man replied. Stunned by this response, Kelly asked him to repeat his answer three times to make sure there had been no mistake or mistranslation.

When Kelly and his colleague Chris Davis began asking more pointed questions about smallpox research at Vector, the Soviet escorts suddenly halted the tour and kept the two British inspectors waiting in an office for more than an hour while they made several frantic telephone calls. Finally, Sandakhchiev and his deputy Netesov agreed to meet with the inspectors in the administration building for a question-and-answer session. The director admitted only that Vector was conducting scientific research with noninfectious cloned fragments of variola DNA that had been furnished by the WHO repository in Moscow. Despite these denials, the U.S. and British inspectors were left with the strong suspicion that Vector was experimenting with the live virus.

Following the U.S.-British inspection, Lev Sandakhchiev sought to develop a cover story to explain the presence of live variola virus at Vector. To this end, he organized an expedition to the Soviet Arctic in the spring of 1991 to recover samples of live virus from the remains of nineteenth-century smallpox victims buried in the permafrost. During smallpox epidemics in the far north, the native peoples had buried their dead in "icehouses," deep pits cut in the ice that were used to store provisions. Because variola virus is extremely stable at low temperatures, it seemed likely that the frozen corpses would still harbor infectious virus particles.

After landing at a remote airstrip north of the Arctic Circle, a team of Vector scientists led by Dr. Alexander Guskov drove for hours across the Siberian tundra. They finally arrived at the village of Chersky on the edge of the Yakutian permafrost. At the site of a smallpox cemetery, they dug trenches ten to twelve feet down through the frozen earth and exposed one of the ice pits. At the bottom lay the 145-year-old remains of several small-pox victims, still wrapped in caribou-skin shrouds. Although the bodies had decomposed, the scientists took biopsies of bone and skin fragments and brought them back to the laboratory for analysis, but no trace of the virus was found.

In July and August 1991, the Vector team returned to the Arctic and ex-cavated another smallpox cemetery. This time the corpses were so well pre-served that the pockmarks were still visible. The scientists took scrapings from the skin lesions to test in the lab. Antibodies against variola virus reacted with the tissue specimens, suggesting that protein remnants of the virus remained, and genetic tests revealed fragments of smallpox DNA. Once again, however, efforts to isolate viable virus particles were unsuccessful. The most likely explanation was that over the decades, the Arctic tundra had undergone repeated cycles of freezing and thawing, causing the virus particles to dis-integrate. Although the Vector scientists could not rule out the possibility that smallpox victims buried elsewhere in the Arctic might still harbor the intact virus, they lacked the funds to conduct additional expeditions.

In December 1991, a Soviet inspection team arrived in the United States to conduct reciprocal inspections of four biological facilities that Soviet mili-tary intelligence claimed were engaged in offensive biowarfare activities. The Soviet team consisted of five high-ranking officers from the Ministry of Defense, two diplomats from the Ministry of Foreign Affairs, four officials from Biopreparat (including Alibekov and Sandakhchiev), one from the Ministry of Health (actually an undercover officer from the GRU, the Soviet military intelligence service), and an interpreter (also from the GRU).

The Soviets had asked to visit four U.S. government biological facilities in various parts of the United States: Fort Detrick, Maryland; Pine Bluff Arsenal, near Little Rock, Arkansas; Dugway Proving Ground in Utah; and the Salk Institute in Swiftwater, Pennsylvania, then the U.S. Army's prime contractor for biodefense vaccines. Some of these sites had been involved in the development, production, or testing of biological weapons during the 1960s, before President Nixon closed down the offensive program. Much to their surprise and frustration, however, the Soviet inspectors were unable to find any evidence of offensive development or production at the four sites. The Army's old biological weapons production facility at Pine Bluff had been abandoned and partly dismantled in 1970–72, and all that was left were decaying buildings, broken fermentation tanks, and rusting railroad tracks. By now, Alibekov was convinced that the United States did not have an active biowarfare program.

In early January 1992, the Soviet inspection team flew home to a new country. Gorbachev had resigned on Christmas Day 1991, when the Soviet Union had officially ceased to exist and had been replaced by fifteen newly independent republics. Symbolizing that dramatic political transformation, the hammer-and-sickle flag had been pulled down from the pole atop the Kremlin and replaced with the Russian tricolor. Moscow was now the capital of the Russian Federation, and Boris Yeltsin was its president. Attitudes, however, had not changed overnight. To justify the continuation of the offensive biowarfare program, the Soviet—now Russian—military brass wanted to persuade President Yeltsin that the United States was secretly retaining "mothballed" biological weapon production facilities that could easily be reactivated. All of the military officers on the inspection team, including Colonel Alibekov, were ordered to fabricate evidence to this effect in their trip reports.

By now thoroughly disillusioned, Alibekov was the only military officer who refused to obey the order. After this act of insubordination, he had no choice but to resign his army commission on January 13, 1992, and a month later he left his position at Biopreparat. Although he soon found work as the Moscow representative of a Kazakhstani bank, he discovered that he was under intensive surveillance: His phone had been tapped and a team of KGB agents was watching his apartment. Fearing for his safety and that of his family, he decided to move back to his native Kazakhstan. Shortly after arriving in Alma-Ata, he was invited by the newly established Kazakhstani government to establish a biological weapons program in that country. Alibekov refused and, with the help of the CIA, made his way to the United States with his family in October 1992. Settling in suburban Virginia, he changed his name to a more American-sounding variant: Kenneth Alibek.

Every weekday for most of the next year, from October 1992 to September 1993, Alibekov was debriefed by two CIA officers in a nondescript office building located in a Virginia suburb near Washington, D.C. Sometimes, experts from other U.S. government agencies participated in the debriefing sessions. Although Alibekov addressed all aspects of the Soviet biological weapons program, he devoted particular attention to smallpox. He described in detail the development program dating back to 1947 and the operational doctrine by which the Soviet military had planned to employ smallpox as a strategic weapon. He also revealed how scientists at Zagorsk and Vector had developed complete processes for manufacturing and storing large quantities of variola virus in stable form, and the intensive effort by Vector scientists to study the molecular biology of variola virus with the aim of enhancing its virulence.

Although Alibekov believed that the twenty-ton stockpile of variola virus at Zagorsk had been destroyed in the late 1980s, he confirmed U.S.

suspicions that seed cultures of the live virus were not restricted to the official WHO repository in Moscow but also existed at Vector and Zagorsk. With respect to the genetic engineering of smallpox, Alibekov knew firsthand only that scientists at Vector had inserted foreign genes into vaccinia. At the time, he had been busy preparing for the U.S.-British inspections of Biopreparat facilities and had lost track of the progress of the research. He admitted that he had no direct proof that the experiments with vaccinia had been repeated successfully with variola major. Still, knowing the scientists' intent and plotting their trajectory, it was within the realm of possibility that Vector had developed a smallpox chimera. After the breakup of the Soviet Union in 1991, however, Vector's research funding had been cut sharply, and none of the genetically engineered strains had been converted into weapons.

In addition to his year-long debriefing by the CIA, Alibekov wrote a lengthy paper on the Soviet biological weapons program that, when completed, was immediately classified at a level above top secret, so that he could no longer read it. No U.S. government officials outside the national security community—including D. A. Henderson—would learn about the vast Soviet effort to turn smallpox into a weapon for another three or four years, and the American public would be kept in the dark until early 1998.

NINE

STAY OF EXECUTION

In May 1990, U.S. Secretary of Health and Human Services (HHS) Louis W. Sullivan, who was totally unaware of the Soviet biowarfare program, gave a speech at the World Health Assembly in which he proposed a joint effort by the United States and the Soviet Union to determine the DNA sequence of representative strains of variola virus. By analyzing the precise order of chemical letters that made up the virus's genetic code, it would be possible to extract the essential scientific information needed to identify smallpox if it ever reemerged in the future.

Sullivan also proposed that once the DNA sequencing work had been completed, both countries would eliminate the collections of variola virus strains stored on their territories in order to ease fears that smallpox might be employed as a biological weapon. "There is no scientific reason not to destroy the remaining stocks of wild virus," he declared. "So I am pleased to announce today that after we complete our sequencing of the small-pox genome, the United States will destroy all remaining virus stocks. I invite our colleagues in the Soviet Union to consider the same course of action. Perhaps we can jointly announce the final elimination of the last traces of this lethal virus."

As part of the warming of U.S.-Soviet relations under Presidents Bush and Gorbachev, Moscow accepted Secretary Sullivan's invitation. In late 1991, the Pentagon and HHS signed a "memorandum of understanding" giving the U.S. Centers for Disease Control and Prevention (CDC) the lead role in the DNA sequencing effort and allocating two years to complete the task, with a view to destroying the U.S. stocks of variola virus by December 31, 1993. Chosen to direct the sequencing project was Dr. Joseph Esposito, who headed the WHO collaborating center on smallpox at the CDC. A compact man with curly black hair, a trimmed black beard, and dark, intelligent eyes, he worked in a cramped cinderblock warren on the CDC's main campus, a cluster of interconnected brick-and-concrete buildings along Clifton Road in Atlanta.

The smallpox virus that Esposito chose to sequence was a virulent strain of variola major called Bangladesh-1975, after its country of origin and year of isolation. Because of the large size and complexity of the viral DNA, it posed a major technical challenge. Even with the use of an automated sequencing machine, the task required more than a year and a half of full-time work by scientists at the CDC and the National Institutes of Health. A computer printout of the finished DNA sequence, consisting of 186,103 chemical units represented by the letters A, C, G, and T, filled 285 loose-leaf pages in eight-point type. This sequence was analogous to an engineer's blueprint of the smallpox virus.

In 1992, after the breakup of the Soviet Union, a team of Vector scientists led by Sergei N. Shchelkunov began sequencing the India-1967 strain of variola major with funding from the WHO. To provide a legitimate basis for this effort, the Russian government issued a special decree creating "The National Program on Conservation of the Genetic Material of the Russian

Collection of Variola Virus Strains." Under this program, Vector scientists would create and characterize a library of cloned DNA fragments from one of the strains of smallpox virus, and then conduct sequencing and computer analysis of the viral genes and the proteins they encoded. It was no coincidence that the strain of variola major the Russians chose to analyze was India-1967—the same strain that the Soviet military had developed into a biological weapon and produced in multiton quantities. Indeed, Vector had already begun secretly analyzing the genetic structure of India-1967 during the late 1980s as a basis for developing the next generation of "improved" smallpox weapons.

In April 1992, Russian President Yeltsin admitted that the Soviet Union had violated the Biological Weapons Convention and issued an edict banning further offensive research and development. As a result of Yeltsin's order and the financial crisis gripping Russia after the breakup of the Soviet Union, the Biopreparat institutes suffered severe cutbacks in government funding. By late 1992, Vector was half empty and protected by a handful of guards who had not been paid in months. Only one-third of the scientific staff was receiving a regular salary, which averaged about $60 a month and was often paid after substantial delays. Money was even lacking to purchase food pellets for the guinea pigs. For this reason, the DNA sequencing work provided a vital source of funding for the laboratory. When the Vector scientists finished sequencing the India-1967 strain in August 1993, they collaborated with the CDC-NIH team to sequence a Brazilian strain of variola minor known as Garcia-1966, and partial sequences of a few other strains.

Some experts argued against publishing the complete DNA sequences of the virulent strains of variola because of concern that rogue states or terrorists might someday attempt to reconstitute them in the laboratory. Although no one had yet succeeded in building an infectious virus from its constituent parts, this feat might become possible as knowledge advanced.

But CDC officials, arguing that scientific inquiry relies on the free and open exchange of information, decided in 1993 to make the DNA sequences publicly available.

The completion of the DNA sequencing project opened the way for the planned destruction of the variola virus stocks in Atlanta and Moscow. At that time, D. A. Henderson was agnostic on the issue of whether or not the virus stocks should be destroyed. Not having access to Alibekov's top-secret revelations, he believed that no research with the live virus had occurred for more than a decade and that none was planned. Under those conditions, it was extremely unlikely that the virus could escape from the CDC repository and, given the Russians' extensive involvement with and apparent pride in the smallpox eradication campaign, it was reasonable to assume that they would keep their own stocks safe and sound. Although it was possible that the virus could have been retained inadvertently by other laboratories, Henderson had given little thought to the intentional development of smallpox as a biological weapon. He was aware, however, that several developing countries that had suffered disproportionately from smallpox were clamoring for the virus stocks to be destroyed. These countries objected to the fact that the WHO had authorized only the United States and Russia to retain samples of the smallpox virus—a policy that had aroused considerable resentment internationally, colored by a degree of paranoia about Washington's and Moscow's intentions.

Meanwhile, a debate over the variola virus stocks was emerging within the scientific community. The executive boards of four leading scientific organizations had endorsed destruction of the virus: the American Society for Microbiology, the International Union of Microbiological Societies, the Board of Counselors of the CDC's National Center for Infectious Diseases, and the Russian Academy of Medical Sciences (although it later reversed its position). Seeking to canvass the views of the broader scientific community on whether or not to destroy the variola virus stocks, the WHO organized a discussion of the issue at the Ninth International Congress of Virology, which was held in August 1993 in Glasgow, Scotland. About five hundred virologists showed up for a special one-hour roundtable session titled "Smallpox: The Final Steps Toward Eradication." This session was structured as a debate between two leading virologists, one favoring destruction of the virus stocks and the other opposed.

The advocates of virus destruction, or "destructionists," included many veterans of the smallpox eradication campaign who had witnessed the ravages of the disease firsthand and considered variola virus to be evil incarnate. For them, the destruction of the virus stocks was the logical culmination of their long, arduous crusade. In the words of Otar Andzhaparidze, director of the Research Institute for Viral Preparations in Moscow, not destroying the virus would be like "playing a piece of music without striking the final note."

From a practical standpoint, now that the DNA sequences of representative strains of variola virus had been determined, the live virus was no longer needed to identify smallpox if it were to reappear in the future. Nor would live variola virus be required to protect against a future outbreak of smallpox, since the smallpox vaccine—based on the distinct vaccinia virus—could be retained and stockpiled for insurance purposes. Destructionists also challenged the claim that it was necessary to keep live variola virus to study the mechanisms of infection. Since smallpox was a uniquely human disease, its eradication meant that clinical studies were now impossible. Although a genetically engineered mouse incorporating human genes might someday be developed, such an animal model would be so remote from the true complexity of the human immune system as to provide few valid scientific insights. Instead, destructionists advocated the study of animal poxviruses, such as mousepox or monkeypox, in their natural hosts. The basic mechanisms by which these viruses caused disease were similar to those of variola, yet they were much safer to handle.

Finally, destructionists argued that eliminating the remaining laboratory stocks of variola virus would strengthen moral suasion against the possible use of smallpox as a biological weapon. As long as the official WHO repositories in Atlanta and Moscow continued to exist, they might provide a pretext for clandestine work on smallpox by "rogue" states seeking biological weapons. Conversely, destroying the official stocks and banning any further possession of the virus would send the clearest possible message to all countries that the military development of smallpox would henceforth be considered a violation of international law, punishable by national and international authorities. As Jeffrey Almond of the University of Reading pointed out, "In an ideal world, I'd argue for the preservation of smallpox for ongoing research. But we don't live in such a place. I think this is an instance in which we need to take the moral high ground and destroy our stocks to send a very clear signal to these other countries that it's a crime against humanity to develop such weaponry."

Advocates of retaining the variola virus stocks, or "retentionists," argued that it was distinctly possible that smallpox might reemerge at some time in the future. Potential sources for a return included the corpses of smallpox victims preserved in Arctic permafrost, samples of variola virus inadvertently or deliberately retained in laboratory freezers, and the possible mutation of monkeypox virus. This virus was known to cause a human disease in West and Central Africa that was clinically almost indistinguishable from smallpox but had a limited potential for person-to-person spread. It was conceivable, however, that random genetic changes in the monkeypox virus might someday give rise to a more virulent and contagious form of human disease, posing a resurgent public health threat. These concerns made it necessary to study the live variola virus in order to be better prepared for such an eventuality.

Retentionists also argued that basic research with live variola virus promised exciting new insights into the process of viral infection and the workings of the human immune system. It was already clear that the mechanisms by which variola virus caused disease were extremely complex, involving the production of multiple viral proteins that interfered with the immune responses of the host. These mechanisms could not be deduced from the viral DNA sequence alone but could only be studied with the intact, infectious virus. According to Dr. Alan Zelicoff, a biologist at Sandia National Laboratories, attempting to understand variola by studying its DNA sequence was like trying to analyze a Swiss watch by smashing it with a hammer and examining the pieces. Although research on smallpox was currently hampered by the lack of a good animal model, scientific advances over the next few decades might well yield a suitable experimental system. For this reason, Zelicoff argued, foregoing the opportunity to obtain valuable new knowledge from variola virus for "political" reasons would be a terrible waste, like throwing money out the window.

Retentionists also expressed concern that if the official repositories of variola virus were destroyed, countries might become complacent and not maintain adequate stocks of smallpox vaccine indefinitely. As for the normative argument, they countered that rogue states and terrorists were not swayed by moral principles and that destroying the known virus stocks would create a false sense of security. "Regrettably, taking the high road in this matter is unlikely to deter unsanctioned behavior; setting good examples rarely prevents criminal behavior," observed Wolfgang Joklik of Duke University Medical School. "The symbolism of destroying the remaining stocks of smallpox virus is highly unlikely to influence anyone contemplating biological warfare or terrorism."

Finally, the two sides disagreed on whether the destruction of the smallpox virus had broader philosophical implications. Retentionists argued that by destroying the variola virus stocks, humans would be deliberately causing the extinction of a living species for the first time in history. However useless and harmful the virus might seem, such an action could set a dangerous precedent. Destructionists countered that when compared with the wholesale extinction of animal and plant species caused by human intervention in natural ecosystems, concern about preserving a deadly virus was misplaced. Moreover, because the DNA sequence of variola virus was extremely similar to that of vaccinia, only a relatively small amount of genetic material would be lost if the smallpox virus disappeared.

In the fall of 1993, the groundswell of scientific opposition to destroying the variola virus stocks, mostly from a small but vociferous group of American virologists, led WHO officials to consider delaying the execution date planned for December 31. Dr. Yuri Ghendon, a Russian virologist at WHO headquarters, told a reporter, "There's no question the virus will be destroyed. The question is, 'Is this the correct time?'"

Another problem was the thorny legal question of who actually "owned" the virus stocks stored in Atlanta and Moscow. Were they just being held in trust for the international community, or did the CDC and Vector have some special legal rights as the physical guardians of the repositories? Was it necessary to obtain permission to destroy the stocks from the countries and organizations that had "loaned" their strain collections to the CDC, namely Great Britain, Japan, the Netherlands, the U.S. Army, and the ATCC? And could the individual countries that had provided medical specimens, such as scabs, legitimately claim title to the viral isolates derived from them? No one had attempted to address these complex legal questions.

In late December 1993, with the destruction deadline only a week away, it was clear that an international consensus did not exist to carry out the execution. Acting CDC Director Walter R. Dowdle told the *New York Times*, "Everyone's feeling is, what's the hurry, give it a little more time and let's talk further about this. . . . The Russians are going through the same process. There are camps that feel the virus should be destroyed and other camps that want to keep it."

Officially, the U.S. government continued to favor the immediate destruction of the virus stocks. In a letter to WHO Director-General Hiroshi Nakajima dated December 23, HHS Assistant Secretary for Health Philip

Lee wrote: "I recognize that postponement beyond December 31, 1993, for destruction of virus stocks may be necessary to permit still wider international scientific discussion. I request, however, that the postponement be time-limited and that an appropriately constituted WHO [scientific advisory committee] be convened . . . to set a final date for destruction."

In Moscow, meanwhile, concern was growing over the safety of the smallpox virus repository at the Research Institute for Viral Preparations. The laboratory had halted all work on smallpox in 1982 to focus on developing a diagnostic test for AIDS, but the variola virus stocks continued to be stored there. Located in an industrial zone on the southeastern edge of the city, the institute was a four-story brick structure painted a drab shade of yellow. The building was deteriorating, its supplies of water and electricity were unreliable, and the security arrangements were limited to an iron fence topped with barbed wire and a small brick guardhouse near the front gate. In a laboratory on the fourth floor of the institute, the Russian stocks of variola virus were stored in two ordinary kitchen freezers behind a pair of glass doors. The doors were locked and wired with special alarms that, in the event of an attempted break-in, would summon the Moscow police from a nearby station. Unlike the CDC, however, the institute did not employ its own armed guards or private security personnel. The smallpox repository was therefore potentially vulnerable to terrorist attack or criminal theft.

During the fall of 1993, the security problems at the Moscow institute became even more acute. On September 21, Russian President Yeltsin, facing strong opposition to his economic and political reforms from the Communist majority in the Supreme Soviet, gave a televised address in which he declared he was dissolving the parliament and ordering new legislative elections. Hours after this announcement, the Supreme Soviet held an emergency session during which Yeltsin's archrival, Vice President Alexander Rutskoi, declared himself president and was sworn in by speaker Ruslan Khasbulatov. About one hundred deputies and several hundred armed supporters occupied the huge white-marble headquarters of the Supreme Soviet, known as the Russian White House, and prepared for a long siege.

On October 2, supporters of the parliament erected barricades and blocked traffic in downtown Moscow, and the following day they attacked a police cordon around the White House and the Moscow city mayor's office. On October 4, after a night of violence in which supporters of Rutskoi and Khasbulatov tried to storm the central Moscow TV station, several elite divisions of the Russian army were mobilized to put down the insurrection. Around 5:00 A.M., tanks rumbled through the streets of Moscow and positioned themselves across from the Russian White House. A few hours later, CNN broadcast dramatic live images of the tanks firing shells into the office tower, and smoke and flames rising from the windows. Meanwhile, near-constant sniper fire from the upper stories of several nearby buildings killed or injured dozens of onlookers. That evening, special army troops finally entered the White House and arrested Rutskoi and Khasbulatov. By then, the official toll from the uprising was 146 dead and about 1,000 injured.

During the violence in Moscow, Svetlana Marennikova, a poxvirologist who had worked for many years at the Research Institute for Viral Preparations, fled to her dacha outside the city, terrified by the possibility that rioters might break into the institute and steal some of the vials containing variola virus. She was unaware that the institute's director, Otar Andzhaparidze, had hired a contingent of private security guards to protect the smallpox repository throughout the uprising. Nevertheless, the ongoing political turmoil in Moscow and the rising crime rate spawned persistent concerns about the security of the virus stocks.

Because of personal conflicts with Andzhaparidze, who had no interest in poxvirus research, Marennikova resigned from her job at the Moscow institute in early 1994 and, although officially retired, became a consultant to Vector. From then on, she and Lev Sandakhchiev lobbied the Russian government to transfer the variola virus stocks to Vector, which had considerably better physical security and containment systems. Possessing the unique strain collection would also enable Vector to become a leading world center for poxvirus research, guaranteeing its scientific and financial future.

On July 29, 1994, the Russian government secretly decided to transfer the variola virus stocks to Vector without informing the WHO in advance. Sealed in hermetic double-walled containers under armed guard, the strain collection was transported by train from Moscow to Koltsovo, arriving at the remote laboratoi y complex on September 30, 1994. In December, the Russian Ministry of Health informed the WHO of the transfer as a fait accompli. WHO officials were shocked and outraged by the Russian action, but they had little recourse as there was no rule against transferring the official virus repository from one laboratory to another.

The variola virus stocks at Vector were stored in a liquid-nitrogen freezer in a vaulted room in Building 1, a multistory building containing several maximum-containment laboratories. To gain access to the repository, one had to pass through a series of submarine doors and then enter a secure area. In addition to electronic sensors and closed-circuit TV cameras, the vault was kept under continual observation by an armed guard who could not open the door without warning other guards and breaking a seal bearing a number and official signatures. Access to the smallpox repository was limited to six scientists who had been personally authorized by Sandakhchiev. No more than two people—a scientist and a security officer—were allowed in the vault at one time. Because the six scientists with access to the virus stocks received a good base salary, as well as funding from national and international research grants, they all had a strong incentive to stay in their jobs.

Kanatjan Alibekov believed that the transfer to Vector of the Russian smallpox virus repository was akin to letting the proverbial fox guard the henhouse. Based on his personal experience, he was convinced that behind a new façade of openness and cooperation, Sandakhchiev and other Vector scientists still sought to identify aspects of variola virus that would be useful for the development of next-generation biological weapons, including host specificity, virulence factors, mechanisms of interaction with the host's immune system, and means for achieving resistance to vaccines and antiviral drugs.

U.S. government officials were also troubled by the evidence that the Soviet Union had worked secretly on smallpox for decades at Vector and Zagorsk, defying the WHO regulations restricting all stocks of variola virus to the official repository in Moscow. Now, by taking a unilateral decision to move the strain collection to Vector, the government of Russia had basically thumbed its nose at the WHO. But because the transfer was not technically illegal and the international organization had no powers of enforcement, nothing could be done.

On September 9, 1994, a group of scientific advisers to the WHO, known as the Ad Hoc Committee on Orthopoxvirus Infections, met in Geneva to discuss the fate of the variola virus stocks. When the ten members of the committee took a vote on the timing of virus destruction, eight of them— including D. A. Henderson—favored destroying the stocks on June 30, 1995. Two members, Keith Dumbell and Svetlana Marennikova, proposed retaining the virus for up to five years to conduct additional research, but they were overruled. After the meeting, committee chairman Frank Fenner said that he understood the excitement generated by recent scientific discoveries about variola virus and the pleas of specialists to postpone destruction. But, he added, "there will always be more to be done and one could keep on deferring the decision forever, and we felt that it should be made now." Fenner predicted that the next World Health Assembly in May 1995 would endorse the committee's recommendation.

In late 1994, with the scheduled destruction of the variola virus stocks less than a year away, the United States government began reviewing its national policy on smallpox in preparation for upcoming decisions by the WHO. One of the U.S. officials involved in the policy review was D. A.

Henderson, who had moved from academia to government service in 1991. After working for two years in the White House science and technology policy office, he had been appointed deputy assistant secretary for health at HHS in 1993.

Now that the DNA of several representative strains of variola virus had been sequenced completely or partially, Henderson believed that the time had finally come to destroy the virus stocks. Everything seemed to be on track for a positive decision on destruction at the upcoming meeting in January 1995 of the WHO Executive Board. This body was a steering committee of thirty-two country representatives, with seats allocated regionally, that developed policy recommendations for the annual May session of the World Health Assembly.

Before the U.S. government could sign off on destruction of the variola virus stocks, however, it was necessary to consult with all the federal departments and agencies that had a stake in the issue. Although the executive branch of the federal government appears monolithic, it is actually quite decentralized: The various departments and agencies are quasi-independent fiefdoms that have distinct policy positions and compete among themselves for high-level political attention, desirable missions, and budgetary resources. As a result, policy development within the U.S. government usually involves a negotiation among the agencies with a stake in a given issue area, and the final policy is often a compromise among competing agency preferences.

On defense and security issues, the job of coordinating the policymaking process falls to the National Security Council (NSC) staff, a small but powerful bureaucracy headed by the president's national security advisor. Most of the officials serving on the NSC staff are on loan from other agencies, such as the State or Defense departments. NSC senior directors chair interagency working groups and guide policy deliberations on a variety of matters affecting national security.

As a health-related topic, the destruction of the variola virus stocks was an "orphan issue" that did not fall neatly into any of the existing portfolios of the NSC staff, such as defense policy, weapons proliferation, or export controls. The issue was therefore assigned by default to Elisa Harris, the NSC official responsible for coordinating U.S. efforts to combat the spread of biological and chemical weapons. A former defense analyst at the Brookings Institution, a Washington think tank, Harris had been appointed to one of the few political slots on the NSC staff at the beginning of the Clinton administration. She had a sharp mind, a driving ambition, and a brusque, arrogant manner.

Harris believed that in order to demonstrate the durability of U.S. policy commitments, it was important to follow through with the long-standing pledge to destroy the variola virus stocks. She also noted that if the official strain collections were destroyed, Washington would gain greater diplomatic leverage in orchestrating political, economic, and military sanctions against any country that chose to retain the virus in defiance of international norms.

On December 20, Harris chaired an interagency meeting on smallpox at the Old Executive Office Building next to the White House. Attending were officials from the Department of State, the Department of Defense (including the Office of the Secretary of Defense, the Joint Chiefs of Staff, and the Defense Intelligence Agency); Health and Human Services (D. A. Henderson and two others); the Department of Commerce, the Arms Control and Disarmament Agency, and the CIA. At the start of the meeting, Harris requested the views of the participating agencies on whether to reaffirm the U.S. commitment to destroy the variola virus stocks on June 30, 1995. All of the agency representatives concurred, with the exception of those from the Department of Defense. These officials argued that it might be necessary to retain live variola virus to produce more vaccine in the future, and that having a library of smallpox strains would help to identify the source of the virus if it was ever deliberately released. When

Henderson pointed out that variola virus was not required to make smallpox vaccine and that the existing strain collection at the CDC was far from
comprehensive, the Pentagon representatives advanced some other ration-
ales for virus retention that had equally little basis in scientific fact.

During his previous stint at the White House Office of Science and
Technology Policy, Henderson had tried to ensure that objective scientific
advice was fed into the higher echelons of policy formulation, and he perceived an excellent opportunity to do so here. Smallpox was a particularly
challenging issue for government policymakers because it was both technical and interdisciplinary, requiring officials to navigate unfamiliar waters.
Accordingly, Henderson proposed that the participating agencies attempt
to arrive at a common understanding of the scientific facts as a starting
point for exploring the policy options related to smallpox virus destruction. His suggestion, however, did not attract much support around the
table. Indeed, one colonel from the Joint Chiefs took umbrage at Henderson's
implication that the Pentagon representatives did not have the necessary
scientific knowledge. He asserted in a booming voice that they possessed
all manner of expertise and that he himself had a doctorate in science—in
political science. This intervention was followed by an embarrassed silence.

The lack of interagency consensus meant that U.S. government policy
on destruction of the variola virus stocks was still unresolved. Although the
State Department and the Arms Control and Disarmament Agency favored
staying the course with the existing policy for political reasons, they deferred
to the Department of Defense on technical matters of biological warfare
defense. Henderson was irritated by the Pentagon's belated intervention
in the smallpox policy debate. Ever since 1991, when Defense had signed a
memorandum of understanding with HHS stipulating that the variola virus
stocks should be destroyed as soon as the DNA sequencing studies were
completed, the military had taken little interest in the issue.

On December 22, 1994, Elisa Harris held an hour-long conference call
with technical advisers from HHS and Defense to discuss the scientific and

national security implications of destroying the variola virus stocks. The participants agreed that the live virus would not be needed to diagnose smallpox were it to reemerge, and that the testing of protective equipment, such as air filters and gas masks, could best be accomplished with surrogate viruses. Nothing was said about developing a new drug therapy or vaccine. The chief argument made for retaining the virus stocks was that they might, in future, permit some scientific studies that could not now be defined. Accordingly, the question for U.S. policy was to balance the desirability of keeping the virus for prospective research purposes against the uncertain risks associated with its retention.

As the smallpox question gained greater salience within the Pentagon bureaucracy, more senior officials became involved, including Franklin C. Miller, the principal deputy assistant secretary of defense for international security policy, and Dr. Stephen Joseph, the assistant secretary of defense for health affairs. Miller, a specialist in nuclear weapons policy, had been deeply alarmed by Alibekov's still-classified revelations that the Soviet Union had turned smallpox into a strategic weapon that could be delivered by intercontinental ballistic missiles against U.S. cities. Just as disturbing was the possibility that, after the breakup of the Soviet Union, seed cultures of the virus could have been smuggled out of Russia to hostile countries, such as Iraq and Iran. If Alibekov's suspicions were correct, then destroying the two official repositories would provide little confidence that the smallpox virus had been eliminated forever from the planet. In Miller's view, the extreme vulnerability of the U.S. civilian population to the disease meant that smallpox posed an existential threat to the United States comparable to that from nuclear weapons.

Joseph, the top Pentagon official responsible for health matters, was a physician by training, although he lacked in-depth knowledge of infectious diseases. He had strong personal views and did not hesitate to express them in an unvarnished and often abrasive manner. Earlier he had sparked a firestorm of controversy by suggesting that "Gulf War syndrome," the cluster of chronic ailments reported by tens of thousands of formerly healthy Persian Gulf War veterans, amounted to little more than malingering. On the smallpox issue, Joseph questioned why the United States had agreed to destroy the variola virus stocks before the medical community had been given a chance to assess whether it could exploit the live virus to develop better defenses against the disease.

Another important adviser to the Pentagon on the smallpox issue was Joshua Lederberg, a Nobel Prize–winning microbiologist and president emeritus of the Rockefeller University in New York. Although Lederberg had worried for decades about the threat of biological warfare, he had long ruled out smallpox as a plausible agent because he could not imagine that a rational actor would employ such an indiscriminate weapon. The contagiousness of the virus meant that it could not be targeted and might well backfire against the attacker's own population. But Alibekov's shocking revelations about the Soviet biowarfare program had forced Lederberg to start taking smallpox seriously as a potential military threat. He was particularly concerned that if a ruthless tyrant like Saddam Hussein had his back to the wall and nothing left to lose, he might consider unleashing smallpox against his enemies as a final instrument of revenge.

During the quiet period between Christmas and New Year's Day, Elisa Harris chaired a classified meeting of government scientists at CIA headquarters in Langley, Virginia, to review the available intelligence on the smallpox issue. During this meeting, several issues were discussed but not fully resolved. Was the existing smallpox vaccine good enough to protect against a military or terrorist attack with variola virus in the form of a concentrated aerosol cloud? Could an adversary defeat the U.S. vaccine by developing genetically engineered strains of smallpox that were more virulent or vaccine-resistant? Was it necessary to sequence additional strains of the virus? Lederberg argued that by sequencing more strains, it would be possible to identify the source of the virus if it were ever used as a biological weapon. Others noted, however, that the CDC strain collection was far from complete and had major geographical and temporal gaps.

With the January 1995 meeting of the WHO Executive Board fast approaching, the time left to reach an agreed U.S. government position was running out. In an effort to influence the decision-making process, agency representatives engaged in various forms of bureaucratic gamesmanship. Although Harris was the point person for smallpox policy, senior officials on both sides of the debate tried to work around the formal lines of authority. Because Miller and Joseph at the Pentagon knew that Harris favored the destruction of the smallpox virus stocks, they did not attend the interagency meetings she chaired and instead tried to influence policy through "back-channel" telephone calls and memos to her superiors on the NSC staff. Meanwhile, D. A. Henderson continued to believe—perhaps naively—that developing a more complete understanding of the underlying science would help to resolve the dispute.

The internal U.S. government debate also spilled over into the international arena. In early January 1995, at a meeting of Western governments in Geneva to discuss ongoing negotiations to strengthen the Biological Weapons Convention, Major Robert Kadlec and other officials from the policy section of the Office of the Secretary of Defense held informal discussions on the smallpox issue with their British, Canadian, and Australian counterparts. Kadlec sought agreement from these countries to delay the decision on smallpox virus destruction planned for the upcoming meeting of the WHO Executive Board. He explained that because of several unresolved technical issues, the U.S. government was still developing its position and needed more time to reach a full interagency agreement.

British defense officials strongly supported the Pentagon's request for a delay. Like the United States, however, the British government was divided: The Ministry of Health favored immediate destruction of the virus stocks, whereas the Ministry of Defense was strongly opposed. Although the Canadian and Australian diplomats with whom Major Kadlec spoke were more skeptical of the Pentagon's assessment of the smallpox threat and the national security rationale for retaining the live virus stocks, they agreed to pass along the U.S. concerns to their respective governments.

On January 18, 1995, the WHO Executive Board convened in Geneva. Australian delegate Tony Adams had planned to take the lead in supporting the recommendation of the WHO Ad Hoc Committee on Orthopoxvirus Infections to proceed with immediate destruction of the variola virus stocks. Before the meeting, however, Graham Pearson, the director of the British Chemical and Biological Defence Establishment at Porton Down, pressured Adams and other delegates to delay consideration of the issue, in part because Washington was still reviewing its position.

The behind-the-scenes lobbying by British and American defense officials had the intended effect. When the smallpox issue came up during the meeting, Executive Board chairman Jesus Kumate of Mexico said that he had been advised to put off consideration of the matter because a consensus on destruction did not yet exist and the question was too important to settle by a majority vote. Although none of the board members objected to the chairman's decision to postpone a decision indefinitely, this action was unprecedented. WHO spokesman Valery Abramov later observed that he could not recall another occasion when the Executive Board had not followed the recommendation of a WHO scientific advisory committee.

When Henderson learned from Tony Adams about the successful lobbying effort by American and British defense officials, he was outraged. In his opinion, the Pentagon had acted improperly and outside official channels to undermine a long-standing U.S. government position. Several countries also expressed unhappiness with Chairman Kumate's decision to defer consideration of the smallpox issue and called for a floor debate at the upcoming World Health Assembly in May. In the end, however, senior WHO officials decided not to press the matter. The assistant director-general told a reporter, "We will follow the mandates of our governing body, which said back off and lay off this for a while." He added that WHO staff members would discuss the matter of smallpox virus destruction informally with representatives of member countries in the hope of reaching a consensus, and that the issue would be sent back to the Executive Board for consideration "at an appropriate time, hopefully within the near future."

In January 1995, D. A. Henderson had written a letter to Stephen Joseph expressing his concern about the Pentagon's strong opposition to destruction of the variola virus stocks, which, in his opinion, did not have a clear scientific rationale. Dr. Joseph's response, dated February 23, said in part: "My point of view stems from the responsibilities vested in me by the department. We must achieve the fullest medical protection of our forces against biological warfare threats consistent with all the realities that impinge on these types of decisions. . . . That protection cannot be compromised by other issues outside of our national security interests."

Frustrated, Henderson met with Elisa Harris and suggested that because the interagency group remained deadlocked on the smallpox issue, it should be sent to President Bill Clinton for a decision. When Harris replied that she wanted the group to keep working toward a government-wide consensus, Henderson countered that there would always be issues on which agencies disagreed, and that the adjudication of such disputes was a key function of the president. "Sometimes effective decision-making requires suppressing one point of view," he said. "In this case, one can either destroy the virus or not. No compromise is possible on that central point." Nevertheless, Harris felt that the issue was not sufficiently ripe to warrant a presidential decision, which would inevitably alienate the losing side.

In an attempt to break the bureaucratic logjam over smallpox policy, senior officials from Defense and HHS agreed to establish a joint scientific working group of experts from outside the federal government to study the technical aspects of the issue. This "Joint Coordinating Group," chaired by Dr. Michael Ascher of the Armed Forces Epidemiology Board, was made up of virologists and public health specialists who advised the Pentagon and the CDC. On April 20 and 21, 1995, the group convened at the U.S. Army Medical Research Institute of Infectious Diseases (USAMRIID), the Army's leading biodefense research facility, headquartered at Fort Detrick in Frederick, Maryland. The purpose of the meeting was to discuss the potential for a natural reemergence or hostile use of smallpox and what types of defensive research with the live virus might be warranted to address those threats.

Defense Department officials believed that if the scientific advisers were briefed about the military threat posed by smallpox, they would be persuaded of the need to retain the variola virus stocks for defensive research and development. Accordingly, the Pentagon arranged for special one-day security clearances for the full panel so they could attend a classified intelligence briefing on smallpox. Analysts from the CIA and the Defense Intelligence Agency delivered the briefing on April 20 in the conference room at USAMRIID. The bottom-line message was that as many as eight countries—some hostile to the United States—might retain undeclared stocks of variola virus. For this reason, the analysts concluded, smallpox posed a serious threat as a military or terrorist weapon.

Overall, the members of the Joint Coordinating Group were not particularly impressed by the intelligence briefing. The briefers seemed to have a poor grasp of the underlying science and were extremely circumspect about the information on which they had based their conclusions. Some of the scientists objected that without seeing the raw data, which had been withheld to protect sensitive intelligence sources and collection techniques, it was impossible to assess the strength of the evidence. The advisers agreed to compromise, however, by recommending a delay in destroying the variola virus stocks, with the understanding that the Pentagon would undertake a short-term, focused research program to develop improved defenses against the potential military or terrorist use of smallpox. With respect to the upcoming World Health Assembly in May 1995, HHS was now prepared to join with the Pentagon and support "a limited postponement" in destroying the variola virus stocks.

On April 26, 1995, HHS and Defense established a joint task force to develop a smallpox research agenda and monitor its progress. This group, co-chaired by D. A. Henderson and Lt. Col. Terry Rauch, included the most knowledgeable poxvirologists from both departments. In a series of meetings over the next few months, the task force identified three research projects: assessing the ability of smallpox vaccine to protect against high-dose military exposures to variola virus, using monkeypox infection in macaques as a model; screening candidate drugs for their ability to inhibit the replication of variola virus, and devising an improved system for detecting and identifying smallpox on the battlefield.

Although Henderson left HHS in the summer of 1995 to return to his position as professor of public health at Johns Hopkins University, he continued to co-chair the joint task force on a part-time basis. He believed that the two departments had made a "gentleman's agreement" that if the planned smallpox research projects were carried out successfully, the United States would be in a position to halt research with live variola virus and approve the prompt destruction of the virus stocks on June 30, 1996. By early December 1995, the studies had been completed on schedule, a remarkable outcome given the unpredictable nature of scientific research. Believing that all work with the live virus had now come to an end, Henderson left on a business trip to Taiwan. On December 13, however, he received an urgent fax at his hotel in Taipei. The message, from an HHS colleague, stated that Stephen Joseph at the Pentagon had demanded additional drug-screening experiments with the live virus. His rationale was that tests with surrogate viruses had not accurately predicted the effectiveness of all four classes of antiviral drugs in inhibiting the replication of variola virus in cell culture. Henderson was outraged that Dr. Joseph had broken what he perceived as a gentleman's agreement.

By the time Henderson returned from Taiwan in late December, the Defense-HHS entente on smallpox research had collapsed into mutual recrimination and the interagency process remained deadlocked. In order to develop a U.S. government position on destruction of the variola virus stocks in time for the January 1996 meeting of the WHO Executive Board, the NSC staff sent the issue to the White House for an eleventh-hour decision. President Clinton carefully reviewed the options and decided that the U.S. representative should support the recommendation of the WHO scientific advisory committee that the virus stocks be destroyed on June 30, 1996. Because of the strongly dissenting opinion from the Pentagon, however, the president agreed to put some flexibility into the negotiating position. He authorized the U.S. representative to join consensus if the other members of the WHO Executive Board voted to postpone destruction for "rational reasons."

During the meeting in Geneva, Tony Adams of Australia suggested postponing destruction of the variola virus stocks for another three years, until June 30, 1999, to permit additional research with the live virus. The other members of the board—including the U.S. representative—quickly lined up behind the Australian proposal, but the Russian delegate agreed to join the consensus only on the condition that the WHO membership would have to vote again in 1999 to confirm the decision before destruction of the virus stocks could proceed. On May 25, 1996, the World Health Assembly endorsed the Executive Board's recommendation of a three-year delay in destroying the virus. The WHO Secretariat planned to use that time to build an international consensus in favor of destruction.

In late 1996, D. A. Henderson first learned of the Soviet effort to turn smallpox into a doomsday weapon. Although he had left HHS in mid-1995 to return to academia, he had retained his security clearance and continued to serve the department in an advisory capacity. During his time in government, Henderson had attended several briefings on biological warfare topics, but Alibekov's revelations about the Soviet smallpox program had been so closely held that they had never been mentioned. Initially, Henderson was skeptical and even incredulous, but as he gradually learned more details over the next several months, the story became more credible—and more appalling. He felt a mixture of shock, depression, and anger over the news, which he experienced as a personal betrayal, particularly by the members of the Vector group. At the same time, he was convinced that the Soviet scientists with whom he had worked closely for years on the WHO eradication campaign, such as epidemiologist Ivan Ladnyi and virologist Svetlana Marennikova, had been unaware of the military program.

WASHINGTON RECONSIDERS

In early 1998, the WHO sent a survey to its 190 member countries requesting their views on the planned destruction of the variola virus stocks on June 30, 1999. Of the seventy-nine countries that responded to the survey, the United States, Great Britain, France, and Italy said they were undecided; Russia said the live virus should be retained indefinitely; and the others favored destruction. The WHO circular triggered a renewed debate over smallpox policy within the U.S. government, with the Pentagon insisting that the Clinton administration reconsider its pro-destruction position.

Public concern over smallpox also became a factor in the U.S. government debate when, on February 25, 1998, Kanatjan Alibekov—now calling himself Ken Alibek—went public with the explosive allegations about the Soviet biological warfare program that he had kept under wraps for the previous six years. In interviews with the ABC News program *Primetime Live*, the *New Yorker* magazine, and the *New York Times*, Alibekov said that he had decided to speak out publicly to help combat the spread of biological weapons and to seek absolution for having developed them. He also alleged that a vestige of the Soviet offensive biological weapons program still existed in Russia under the cover of defensive research. The Russians, he claimed, sought to preserve their "military biological potential. They keep safe their personnel, their scientific knowledge. And they still have production capability."

Meanwhile, President Clinton was increasingly preoccupied by the specter of biological terrorism. On February 26, 1993, only a month after he had taken office, the United States had experienced the first major incident of international terrorism on American soil when a group of Islamist sympathizers, led by Ramzi Yousef, exploded a powerful bomb under the World Trade Center in New York, killing six people and injuring more than a thousand. Two years later, on April 19, 1995, the bombing of the federal building in Oklahoma City by right-wing extremists Timothy McVeigh and Terry Nichols killed 168 people, demonstrating that domestic terrorists were capable of inflicting large-scale casualties. A third watershed event occurred on March 20, 1995, when a Japanese cult called Aum Shinrikyo released sarin nerve gas on the Tokyo subway, killing twelve people and injuring more than a thousand. Flush with nearly $1 billion in cash from a variety of business enterprises, the quasi-Buddhist group had recruited chemists and biologists from leading Japanese universities to produce a frightening toxic arsenal. Their goals included killing millions, fulfilling the apocalyptic predictions of cult leader Shoko Asahara, and seizing control of the Japanese government.

After the nerve gas attack on the Tokyo subway, a Japanese police investigation revealed that members of Aum Shinrikyo had released two biowarfare agents, anthrax bacteria and botulinum toxin, in downtown Tokyo on nine occasions in 1990 and 1993 with the intent of inflicting mass casualties. Fortunately, technical problems foiled all nine attacks, which caused no known injuries or deaths. Although it was unclear whether the Japanese doomsday cult was a bizarre aberration or the harbinger of a deadly new trend in terrorism, U.S. cities were clearly vulnerable to attack. Accordingly, the Clinton administration began to view bioterrorism as a major threat to national security.

The president's concerns were heightened when he read a science-fiction thriller by Richard Preston titled *The Cobra Event*, published in November 1997. Reveling in gruesome description, the novel depicts a deranged scientist who develops a genetically engineered virus as a weapon of mass destruction. The fictional "Cobra" virus combines the worst attributes of smallpox, the common cold, a rapidly proliferating insect virus, and the gene for Lesch-Nyhan syndrome, a neurological disorder that causes victims to mutilate themselves. After the mad scientist releases the engineered virus into the New York City subway system, ordinary citizens begin to die horrible deaths.

Clinton was so alarmed by Preston's novel that he asked U.S. intelligence analysts to assess its credibility and urged senior administration officials and House Speaker Newt Gingrich to read it. Although government scientists were skeptical that a recombinant virus like the one depicted in *The Cobra Event* would have any likelihood of working, the novel reinforced Clinton's worst fears that terrorists might unleash deadly microbes against American cities. Another scare came in February 1998, when Larry Wayne Harris, an Ohio laboratory technician with white-supremacist beliefs, threatened to release "military grade anthrax" in downtown Las Vegas. Although an analysis of Harris's purported anthrax weapon later revealed that it was a harmless veterinary vaccine strain, sensational media coverage of the Las Vegas incident turned Harris into a poster boy for the emerging threat of bioterrorism.

President Clinton's personal preoccupation with the issue became the driving force behind a series of secret federal meetings, exercises, and presidential directives aimed at strengthening the nation's counterterrorist posture. On a sunny spring morning in March 1998, about forty cabinet members and senior officials from more than a dozen U.S. federal agencies gathered at Blair House, the official guest residence across from the White

House, for a secret exercise to test their readiness to respond to several types of terrorist attacks. One scenario, inspired either by *The Cobra Event* or by Alibekov's revelations, had terrorists covertly release a genetically engineered hybrid of smallpox and Marburg hemorrhagic fever along the U.S.-Mexican border, causing hundreds of thousands of casualties throughout the Southwest. During the simulated outbreak, panic and confusion were widespread and serious problems developed with logistics, medical care, and turf battles among state and federal agencies. The fictional crisis even acquired an international dimension when Mexico threatened military action to keep U.S. citizens from fleeing south across the border to escape the epidemic. Although some aspects of the scenario were clearly implausible, the results of the exercise suggested that a bioterrorist attack with a contagious agent like smallpox could rapidly spiral out of control.

In the aftermath of the Blair House exercise, President Clinton asked the NSC staff to convene a panel of scientific experts to discuss the threat of bioterrorism and how the federal government should respond. The briefing was held on April 10, 1998, in the Cabinet Room of the White House and was moderated by Dr. Frank Young, a Presbyterian minister and former director of the Office of Emergency Preparedness, the HHS office responsible for medical disaster response. The seven-member panel told President Clinton and other senior officials that the threat of bioterrorism was real and criticized the lack of a coordinated federal government plan for dealing with it. Clinton asked a series of probing questions and became so deeply engaged in the discussion that the meeting ran nearly twice the allotted time.

As a follow-up to the briefing, the president asked Dr. Young to provide a written list of recommendations on how to improve the nation's ability to defend against bioterrorism. In a sixteen-page letter report delivered to the White House on May 6, the scientific panel proposed stockpiling vaccines against the most likely biological threat agents, including smallpox, and strengthening the nation's public health system. President

Clinton directed the executive branch to prepare plans and budgets for responding to the panel's near-term recommendations, about three-quarters of which were adopted as U.S. government policies.

On May 22, 1998, only a month after the White House briefing, the president gave a commencement speech at the U.S. Naval Academy in Annapolis, Maryland, during which he unveiled a new Initiative on Biological and Chemical Weapons Preparedness. He told the graduating cadets that because the "enemies of peace" could no longer hope to defeat the United States by conventional military means, they were developing new forms of attack with chemical, biological, and nuclear weapons. To help address this new threat, he named NSC official Richard A. Clarke as the national coordinator for counterterrorism, responsible for overseeing all such efforts by federal government agencies and the military. In addition, Clinton called for stockpiling vaccines and antibotics and upgrading the nation's disease surveillance systems so that they could quickly detect an epidemic caused by the covert release of a biowarfare agent. The Public Health Service was assigned the lead role for managing the medical consequences of a bioterrorist attack.

Observers expressed surprise at how quickly the president and his NSC staff had developed the bioterrorism initiative, noting with some concern that it had not gone through the usual deliberative planning process. Although Congress was favorably inclined toward the program, criticisms began to surface. The New York Times reported that a "quiet war" had broken out within the administration over the president's initiative, with HHS officials contending that it did not give sufficient emphasis to improving the nation's public health infrastructure. A few outside critics also claimed that the White House had exaggerated the threat of bioterrorism. In their view, the failure of Aum Shinrikyo to harm anyone in nine attempted attacks with anthrax bacteria and botulinum toxin demonstrated that biological agents were actually quite difficult for terrorists to produce and deliver effectively.

The perceived salience of the bioterrorist threat increased again, however, on August 7, 1998, when terrorists linked to Saudi fugitive Osama bin Laden bombed the U.S. embassies in Kenya and Tanzania, killing 224 people including twelve Americans. President Clinton later revealed in a newspaper interview that bin Laden had "made an effort to get chemical weapons" and "may have" tried to acquire biological weapons. The administration's worst-case scenario was that hostile states such as Iraq would supply biological agents to terrorists such as Hamas, Hizbollah, Abu Nidal, or bin Laden's group Al Qaeda (Arabic for "The Base") and then step back and deny responsibility. Another plausible threat of biological attack might come from wealthy fringe groups similar to Aum Shinrikyo.

D. A. Henderson shared President Clinton's concern about the threat of bioterrorism, particularly with respect to smallpox. In the years immediately following smallpox eradication, Henderson had considered the odds of an accidental release or deliberate use of variola virus to be vanishingly small. But Alibekov's stunning revelations about the Soviet program had changed that assessment and made the possibility that the virus might fall into terrorist hands suddenly loom much larger. Henderson worried that variola major would make a "good" terrorist weapon because it was hardy and caused a frightening illness with a case fatality rate of 30 percent. Unlike noncontagious agents, such as anthrax or botulinum toxin, smallpox would give rise to an expanding circle of secondary cases, requiring immediate containment by isolation and immunization.

The world's population had also grown strikingly vulnerable to the disease. Routine vaccination of civilians in the United States had stopped in 1972 and in other parts of the world by 1984 at the latest. Because the protective immunity induced by the vaccine lasted only about seven to ten years, it had long since waned for those vaccinated as children, and millions more had never been vaccinated. No effective drug treatments for smallpox were available, and stocks of the vaccine had dwindled around the world. Given these facts, if terrorists were to release an aerosol of variola virus in a densely populated area, the consequences would be catastrophic. Even an outbreak with as few as fifty to one hundred primary cases could result in widespread panic and require large-scale—and perhaps national—emergency control measures. Henderson conveyed his growing concerns about the smallpox threat in a series of articles in scientific journals and speeches at professional meetings.

Because the WHO Executive Board would meet again in January 1999 to discuss the destruction of the variola virus stocks, the U.S. government faced a tight deadline for decision. The policymaking process remained adrift, however, until Dr. Kenneth Bernard joined the NSC staff. A short, intense man in his early fifties with a restless energy, Bernard had received his M.D. from the University of California at Davis in 1975 and a master's degree in public health from the London School of Hygiene and Tropical Medicine. He had spent most of his government career as a physician-epidemiologist with the Public Health Service, including stints with the Peace Corps, the CDC, and the Office of International Health at HHS. His most recent assignment had taken him to Geneva for four years as health attaché at the U.S. mission to the United Nations, where he had served as the U.S. government's liaison to the WHO and had gained valuable diplomatic experience.

During his years in Geneva, Bernard had become increasingly convinced that the security implications of infectious diseases were not receiving the attention they deserved in Washington. Although bioterrorism had put public health on the national security agenda for the first time, it was only a small piece of a much larger puzzle. Bernard was convinced, for example, that the infection of a quarter of Africa's population with the HIV/AIDS virus would lead to widespread political instability and conflict on that continent, with inevitable security ramifications for the United States.

After receiving a promotion to the senior ranks of the Public Health Service, Bernard arranged to meet with HHS Secretary Donna Shalala.

Accompanied by Jo Ivey Bouffard, a trusted Shalala advisor on international health, he proposed creating a new position on the NSC staff for a special assistant to the president for international health affairs. Secretary Shalala was convinced by his arguments and in turn persuaded Sandy Berger, the president's national security advisor, to approve the new position. Thus, in August 1998, Bernard became the first holder of the job he had invented. His initial portfolio covered a wide range of issues bridging international health and U.S. national security, including bioterrorism, the spread of drug-resistant tuberculosis, the impact of AIDS in the developing world, the safety of the U.S. food supply, emerging infectious diseases such as Ebola hemorrhagic fever, the deliberate denial of health care as an instrument of war, and the use of immunization campaigns to reinforce truces in armed conflicts. Bernard also had ambitious personal objectives: His primary goal was to institutionalize the emerging perception of public health as a national security issue so that his position on the NSC staff would be retained by future administrations.

Because Bernard had entered the shadowy world of politico-military affairs from the field of public health, which stressed the free and open exchange of information, his learning curve on the NSC staff was steep and treacherous. He was not prepared for the White House bureaucratic culture of secrecy, intrigue, and interoffice rivalry. Even if one was used to dealing with difficult people, it was hard to adjust to an environment where officials viewed information as the currency of power and refused to share it. The smallpox issue, however, was well suited to Bernard's scientific expertise and diplomatic experience, and it was a natural fit with his new NSC portfolio. Unlike Elisa Harris, he was not identified with one side of the destruction debate and could therefore serve as an honest broker to move the interagency deliberations forward.

In late August 1998, when Harris went on maternity leave to have her second child, Bernard assumed full responsibility for the smallpox issue.

He moved quickly to formalize and centralize the U.S. government policy-making process, which Harris had conducted on an ad hoc basis or had delegated to an interagency group on biological arms control. To engage senior government officials with the power to make decisions, Bernard established a formal Interagency Working Group (IWG, pronounced "eye-wig") on smallpox, with membership at the assistant-secretary level. All agencies with a demonstrable institutional stake in the issue were invited to participate, including HHS, the Office of the Secretary of Defense, the Joint Chiefs of Staff, the Department of State, the Department of Energy, the Arms Control and Disarmament Agency, the Office of Science and Technology Policy, and the intelligence community. In an effort to be inclusive, Bernard solicited the views of bureaucratic players who had been left out of the earlier debate, such as the directors of the CDC, the National Science Foundation, and the National Institutes of Health. He also tried to keep the policymaking process as transparent as possible so that all the stakeholding agencies were aware of the state of play.

Given the large number of unresolved issues related to smallpox research, the Clinton administration decided to commission a study of the future scientific needs for live variola virus by a blue-ribbon panel of experts under the auspices of the Institute of Medicine (IOM), a branch of the National Academy of Sciences. In discussions with the IOM staff in late August 1998, the U.S. government sponsors—HHS, Defense, and Energy—made it clear that they wanted to limit the scope of the study to a technical assessment of what research might be conducted with live variola virus for case detection, diagnosis, prevention, and treatment. Explicitly excluded from the expert committee's mandate was any consideration of costs, relative priorities, or political factors. The rationale for this narrow focus was that other government agencies would be asked to prepare papers analyzing the smallpox issue from the perspectives of intelligence, public health, security, foreign policy, and international law. Senior policymakers would then consider all of these elements when reaching a final decision.

In selecting the members of the expert committee, the IOM staff tried to make sure that the group would appear as objective and balanced as possible. Accordingly, they deliberately excluded scientists who were directly involved in smallpox research or had taken strong positions on either side of the destruction debate. Members were chosen instead for their expertise and academic standing "one step removed" from smallpox. The nineteen scientists selected for the study were mostly virologists, along with a few specialists in infectious diseases, epidemiology, and public health, and one bioethicist. Chairing the committee was Dr. Charles Carpenter, a professor of medicine at Brown University who had headed previous IOM studies and shown himself to be an effective leader and an evenhanded moderator.

To generate useful background information for the committee and the government sponsors, the IOM held a one-day public workshop on November 20, 1998, at which scientific experts on smallpox were invited to give presentations. The meeting took place in the lecture room of the U.S. National Academy of Sciences, a marble building resembling a Roman temple that overlooks the Washington Mall and the Lincoln Memorial. Much of the liveliest discussion during the workshop focused on whether live variola virus would be needed to develop and test new therapies against smallpox.

Dr. John W. Huggins, the director of antiviral drug development at the U.S. Army Medical Research Institute of Infectious Diseases (USAMRIID) at Fort Detrick, gave a presentation on his efforts to screen drugs for anti-smallpox activity, including some preliminary results with a drug called cidofovir. Huggins challenged the assumption that one could rely exclusively on vaccination to contain a future smallpox outbreak. In a highly susceptible population such as that of the United States in the late 1990s, he argued, a single case of smallpox might infect as many as twenty others, each of whom could infect twenty more, so that an epidemic would spread like wildfire. Although containing small outbreaks by ring vaccination had worked well in developing countries during the WHO eradication campaign, the population of an American city was vastly larger and more mobile than that of an African village. It simply wasn't possible to order the inhabitants of New York City to stay holed up in their apartments for two weeks while health officials vaccinated around them.

Huggins claimed that the first wave of cases infected by a terrorist release of variola virus, perhaps numbering in the hundreds or thousands, would probably be acutely ill when the outbreak was detected, at which time it would be too late to vaccinate. He also noted that an increasing number of people with impaired immune function were at risk of serious complications from the smallpox vaccine, including organ transplant recipients, cancer patients on chemotherapy, and those infected with the HIV/AIDS virus.

A case study was suggestive of the problem. In April 1984, a seemingly healthy nineteen-year-old man entered basic training with the U.S. Army and received a battery of vaccinations, including one against smallpox. Two weeks later, he developed fever, headache, and night sweats and was hospitalized with a diagnosis of meningitis. A large skin ulcer appeared at the smallpox vaccination site, followed by pustular lesions all over his body. It turned out that the young man had been infected with HIV before his enlistment and the multiple vaccinations had triggered a full-blown case of AIDS. His crippled immune system could not keep the vaccinia virus in check, and it spread throughout his body.

This tragic case suggested that primary smallpox vaccination of persons with subclinical HIV infection posed a serious risk of vaccine-induced disease. Huggins argued that a decision not to develop an anti-smallpox drug would be tantamount to telling the large group of immunosuppressed people that they would be excluded from the only possible treatment option in the event of a major outbreak. "I think that therapy of existing cases is something that most Americans would assume the government would be willing to do," he concluded. "We're just not going to write off everybody who becomes ill and say, sorry, the vaccine didn't work."

During the discussion, D. A. Henderson expressed doubt about the value of antiviral drugs in dealing with a future smallpox epidemic. He noted that cidofovir had toxic side effects, and that studies using the drug to treat monkeypox in macaques had suggested that it would not be curative once the acute symptoms of smallpox had developed. More generally, because smallpox had been eradicated from the human population and a realistic animal model of the disease did not exist, it would be impossible to validate the effectiveness of any anti-smallpox drug until it was actually used. Instead of developing new drugs, Henderson called for a crash program to produce a large reserve stockpile of smallpox vaccine. Since the efficacy of the vaccine had been demonstrated during the WHO eradication campaign, producing more of the same type would not require retaining the variola virus stocks, which could be destroyed on schedule in June 1999. As for the risks of vaccination in HIV-infected individuals, Henderson believed that the danger was confined to the small minority of patients with severe immune suppression, whose life expectancy was limited in any case. He was, however, prepared to support the development of drugs to control progressive vaccinia, the chief complication of the existing smallpox vaccine.

The IOM workshop also addressed the potential scientific benefits of research with live variola virus. Virologist Wolfgang Joklik argued that because certain proteins produced by the virus were exquisitely adapted to the human immune system and behaved as powerful regulators of immune function, variola was a "Rosetta Stone" for understanding the mechanisms of viral infection. Again, Henderson was skeptical, given the lack of a realistic animal model of the human immune system. Other participants, however, were more optimistic about the prospects for developing such an animal model in the future. The day-long workshop concluded with relatively few areas of agreement.

* * *

In December 1998, an intelligence assessment of the smallpox threat, which Elisa Harris had requested several months earlier from the CIA's Non-Proliferation Center, was presented to the Interagency Working Group in both written and oral forms. Classified at the "secret" level, this assessment summarized a large number of anecdotal reports about undeclared stocks of smallpox virus. The information was strictly circumstantial, but the weight of the evidence suggested that at least three countries—Russia, North Korea, and Iraq—might retain clandestine stocks of variola virus for military use. These disturbing findings soon became public knowledge. Much to the irritation of the U.S. intelligence community, the secret report was leaked six months later to the *New York Times,* which summarized its contents in a front-page story.

Given its long and checkered history with smallpox, Russia was an obvious suspect. The CIA report gave considerable weight to Alibekov's allegations that the Soviet Union had produced variola virus in multiton quantities as a strategic weapon. Although the huge stockpile of the virus had almost certainly been destroyed, it was reasonable to doubt that every last gram had been accounted for. The fate of the seed stocks, production documents, and manufacturing equipment remained unknown, and significant discrepancies existed in the number of smallpox strains that the Russians had reported possessing at various times. Because the amount of "loose change" was considerable, it did not appear to be simply an honest accounting error.

Important questions also remained about the facilities that Alibekov had implicated in offensive military work on smallpox. The Vector laboratory was now fairly open to the outside world and hosted foreign scientific delegations almost continuously, but the Center of Virology in Zagorsk (now renamed Sergiev Posad) remained a top-secret facility operated by the Russian Ministry of Defense, subsidized by the state, and strictly off-limits to foreigners. There was also lingering concern about the "brain drain" of unemployed former bioweapons scientists from Russia. The *New York Times* reported in December 1998 that at least five former Soviet bioweapons scientists were working in Iran, which was paying them $5,000 a month in lieu of their previous $100 monthly salary. Other former Soviet experts might be sharing their deadly expertise over the Internet without ever leaving home. In addition, carrying a few vials containing freeze-dried cultures of variola virus out of Russia in a shirt pocket would be far easier than smuggling radioactive bomb materials or sarin-filled artillery shells. In an interview with CBS News, Alibekov observed, "Thousands of people know how to work with the smallpox virus. Where these people are and what they're doing of course nobody knows. It makes me nervous."

Circumstantial evidence suggested that clandestine stocks of variola virus might exist in other countries as well. In 1993, the Russian Foreign Intelligence Service, the successor to the Soviet KGB, had published an unclassified report titled *A New Challenge After the Cold War: Proliferation of Weapons of Mass Destruction.* The entry for North Korea stated that Pyongyang was "performing applied military-biological research at a whole series of universities, medical institutes, and specialized research institutes. Work is being performed at these research centers with pathogens for malignant anthrax, cholera, bubonic plague, and smallpox." Supporting this allegation to some extent was the fact that blood samples drawn from North Korean soldiers who had defected to the South contained antibodies to smallpox vaccine, and some of the vaccinations appeared fresh. Pyongyang could have acquired strains of variola virus from several sources: seed cultures transferred from the former Soviet Union or China, indigenous stocks dating back to the Korean War, or samples collected during smallpox epidemics in other developing countries during the 1960s and 1970s.

Several pieces of circumstantial evidence suggested that Iraq, which had admitted to having produced anthrax bacteria and other biowarfare agents prior to the 1991 Persian Gulf War, might have a secret stockpile of variola virus. The first clue was Iraqi production and use of smallpox vaccine. According to a declassified report from the U.S. Defense Intelligence Agency (DIA), of seventy-one blood samples taken from Iraqi prisoners during the Gulf War, eight had shown evidence of smallpox vaccination. Inspectors from the United Nations Special Commission (UNSCOM), which conducted inspections of suspected Iraqi biological warfare facilities from April 1991 to November 1998, determined that Iraq had manufactured smallpox vaccine as late as 1989 and was still immunizing troops in 1990, more than decade after the disease had been eradicated.

On its face, the Iraqi immunization of soldiers against smallpox was not necessarily incriminating, since the U.S. military had continued doing so until 1990. But suspicions were further aroused when Hazem Ali, a senior virologist involved in the Iraqi bioweapons program, admitted to having studied camelpox virus, which causes fever and skin rash in camels. Ali claimed that because the Arabs had lived with camels for generations, they were immune to camelpox, whereas Westerners were not. For this reason, he explained, Iraq had developed camelpox as an "ethnic weapon" for selective attacks against foreign troops. In fact, camelpox rarely infects humans, and when it does, the resulting illness is relatively mild. According to Israeli sources of unknown reliability, Iraqi scientists may have attempted to develop a more virulent strain of camelpox, which they reportedly tested on Kurdish prisoners who were forced to serve as human guinea pigs.

Another possible explanation for Iraqi research with camelpox virus was that it was being used as a surrogate for variola. The last smallpox epidemic in Iraq had occurred in 1972, and it was likely that Iraqi scientists had collected isolates of the virus and put them away in freezers for safekeeping. Although camelpox virus is closely related to variola in its genetic structure and physical properties, it is much safer to work with and does not require a high level of physical containment. Thus, if Iraqi scientists planned to develop smallpox as a biological weapon, they could use camelpox to refine their production techniques and to study methods for dispersing the virus.

Further evidence for Iraqi military interest in smallpox surfaced in the mid-1990s, when UNSCOM inspectors visited the maintenance shop of the State Establishment for Medical Appliances Marketing, a branch of the Iraqi Ministry of Health that had been implicated in the biological warfare program. During an inspection of the site, the UN inspectors found an old freeze-drier labeled "smallpox" in Arabic. Although the Iraqis claimed that this piece of equipment had been employed for the production of freeze-dried smallpox vaccine, it could have just as easily have been used to convert a liquid suspension of variola virus into powder form for dissemination as a fine-particle aerosol.

There was also a possible Soviet connection. A declassified DIA intelligence report from May 1994, citing a source of unknown reliability, stated that virologists from one of the Soviet Ministry of Defense's military microbiology institutes had transferred smallpox virus cultures to Iraq in the late 1980s or early 1990s and had assisted the Iraqis with their biological warfare program. In addition, the British Secret Intelligence Service (MI-6) had reportedly spotted a Vector scientist in Baghdad in 1991.

Although the evidence for undeclared stocks of variola virus in Russia, North Korea, and Iraq was almost entirely circumstantial, the members of the Interagency Working Group could not simply dismiss it. Additional leaks and rumors of uncertain reliability suggested that several other countries might have inadvertently or deliberately retained specimens of the virus from the time when smallpox was a common disease. Possible suspects included China, Cuba, India, Iran, Israel, Pakistan, and Yugoslavia.

* * *

Given the likely existence of undeclared stocks of variola virus, Pentagon officials argued that for the WHO to destroy the two official repositories in the United States and Russia and then declare that the virus had been banished from the planet would be "perpetrating a fraud." A few virologists also speculated that a smallpox-like virus might someday be reconstituted in the laboratory. The DNA sequences for variola, vaccinia, monkeypox, and cowpox had all been published in the open scientific literature. Thus, as genetic technology advanced, it might become possible to synthesize smallpox genes and insert them into vaccinia or an animal poxvirus, such as monkeypox or camelpox, to create a hybrid strain that was deadly in humans.

Although D. A. Henderson did not deny that some countries might possess covert stocks of variola virus, he refused to accept this possibility as a valid reason to retain the two official repositories indefinitely. The WHO Ad Hoc Committee on Orthopoxvirus Infections had already examined some hypothetical scenarios for a future reemergence of smallpox, including the mutation of monkeypox virus to become more contagious and deadly in humans, and the possibility that global warming might expose the frozen remains of smallpox victims buried in the Arctic permafrost. In each case, the committee had concluded that the resulting outbreaks would be small and easily containable with vaccination.

With respect to defenses against the deliberate use of variola virus as a military or terrorist weapon, Henderson believed that useful scientific work could proceed without access to the live virus (for example, by using cloned smallpox DNA or surrogate poxviruses) and that destruction of the declared stocks would make a moral statement that even terrorists could not ignore. In his view, Pentagon officials were stuck in the Cold War mindset that as long as other countries had smallpox, the United States had to keep some of its own. Because Washington had no intention of retaliating in kind if it was ever attacked with smallpox, however, the logic of nuclear deterrence did not hold. Instead, Henderson argued that the World Health Assembly should set a date for the destruction of all known stocks of variola virus and declare that any retention of live virus after that point would be considered a grave violation of international law. Once possession had been universally outlawed, the threat of severe punishment against any rogue states, laboratories, or scientists retaining clandestine stocks of the virus would likely diminish the risk of its use as a weapon.

The IOM committee on the future scientific needs for live variola virus met for the second and final time on January 6, 1999, to complete its report. After extensive and sometimes heated debate, the experts managed to reach consensus on all of the major scientific issues within their purview. Because the study mandate explicitly precluded the committee from taking a position on destruction of the virus stocks, the participants sought a way to present their findings and recommendations in a neutral and objective manner. They finally decided to frame their conclusions in conditional terms: *If* the U.S. government wished to develop such-and-such a capability, *then* it would or would not require access to the live virus. The expert committee also subjected every sentence of the one hundred-page report to careful scrutiny to eliminate any possible perception of bias. Following standard IOM procedure, the draft was sent to nineteen outside experts for technical review.

On January 22, 1999, President Clinton gave a speech at the National Academy of Sciences "on keeping America secure for the twenty-first century," in which the threat of bioterrorism featured prominently. "Four years ago," he said, "the world received a wake-up call when a group unleashed a deadly chemical weapon, nerve gas, in the Tokyo subway. We have to be ready for the possibility that such a group will obtain biological weapons. . . . If we prepare to defend against these emerging threats, we will show terrorists that assaults on America will accomplish nothing but their own downfall." The president also told the *New York Times* that in formulating the administration's policy on smallpox, he would rely to a large extent on the forthcoming IOM report, which was due to be released in March. His goal, Clinton said, was to ensure that the United States had "whatever it takes to buy us a response" to a potential reemergence of the disease.

ELEVEN

DECISION IN GENEVA

On March 15, 1999, the Institute of Medicine released its long-awaited report. The key finding was that the most compelling reason for retaining the stocks of live variola virus was to develop antiviral drugs for treating civilian casualties in the event of a major smallpox outbreak, such as might result from a terrorist attack. Since the disease would probably spread widely before large-scale vaccination could begin, anti-smallpox drugs would be needed to treat the first generation of cases and to help contain the epidemic.

The IOM committee also found that access to variola virus would be required "if certain approaches to the development of novel types of smallpox vaccine were pursued." Because the existing vaccine was risky to use in people with weakened immune systems, the committee suggested the possibility of developing a new vaccine that was not based on a live virus like vaccinia. Testing the effectiveness of such a next-generation vaccine would be problematic, however, because of the lack of an animal model of smallpox. For all other biodefense applications, access to the live virus was not essential and could be replaced by surrogate poxviruses and cloned smallpox DNA. The report also noted that "much scientific information, particularly concerning the human immune system, could be learned through experimentation with live variola virus."

The Pentagon was delighted with the findings of the IOM study. Defenders of the report praised its evenhandedness and argued that if the expert committee had tried to assess the political drawbacks of retaining the virus, the integrity of the scientific assessment would have suffered. D. A. Henderson, however, condemned the study as biased and unrealistic. He complained that most of the committee members were academic virologists with no practical understanding of the extraordinarily high costs, technical complexities, and years of work required for drug or vaccine development. Moreover, the IOM committee had been instructed to assess the benefits of scientific research with live variola virus without respect to cost, feasibility, or any other realistic constraints or priorities. In Henderson's view, this overly narrow mandate had been a clear attempt to stack the deck. He even claimed that a member of the IOM committee had privately admitted as much, saying, "We soon knew that it was a setup, but there really wasn't much we could do about it."

Press coverage of the IOM report muddied the waters further by making the committee's policy recommendations appear more definitive than they really were. The scientific experts, carefully following the guidelines laid down by the U.S. government sponsors, had restricted themselves to analyzing the technical issues and had gone to great lengths to avoid taking a position on the fate of the variola virus stocks. Yet print and television reporters, attempting to summarize the study in a few attention-grabbing words, glossed over the caveats and conditional wordings in the IOM report and interpreted the committee's findings as a clear endorsement of virus retention. In a radio interview, Dr. Carpenter expressed his dismay over the media coverage. He confessed that he personally favored destruction of the virus but that the expert committee's conclusions had been dictated by its limited mandate. "We were asked to look at the potential utility of maintaining the live stocks of smallpox virus," he said. "We were not asked to look at the downside of retaining it."

Despite Dr. Carpenter's qualms about the way the IOM report had been interpreted, the advocates of smallpox virus retention within the U.S. government did not hesitate to use the expert committee's findings to buttress their position in the interagency debate. "There's no doubt the report will have a major impact," an unnamed senior administration official told the *New York Times*. "Science is science, and the science that has now spoken will dramatically affect the decision-making process here."

In late March 1999, the Interagency Working Group on smallpox finally reached a unanimous decision. All of the agency representatives agreed that at the upcoming World Health Assembly in May, the United States should seek to postpone the destruction of the variola virus stocks stored at the CDC to permit further defensive research with the live virus. The main factor contributing to this decision was the growing concern over the potential use of smallpox as a military or terrorist weapon. This perception had been strengthened by Alibekov's revelations about the Soviet biological warfare program and the circumstantial evidence for the existence of undeclared stocks of variola virus in Russia, North Korea, and Iraq.

The IOM report also had a major impact on the interagency decision. Of all the scientific arguments for retaining the variola virus stocks, the most persuasive was that the development of anti-smallpox drugs and their approval by the U.S. Food and Drug Administration (FDA) would almost certainly require testing candidate compounds against live variola virus, both in cell culture and in laboratory animals (if a suitable animal model could be found). Senior officials such as HHS Assistant Secretary Margaret Hamburg concluded that if there was real reason to believe that smallpox might be used as a terrorist weapon, it was desirable for the U.S. government to have more and better tools in its defensive armamentarium. To this end, further screening of potential antiviral drugs would be a prudent insurance policy. The other main argument for retention—that basic research on variola virus might lead eventually to scientific and medical breakthroughs—was not compelling enough to win the day on its own.

Because the IWG was proposing a 180–degree change in U.S. policy, the interagency recommendation had to go to President Clinton, who decided to approve it. As head of state, he could not forgo the development of new measures that might protect the U.S. population against a catastrophic incident of bioterrorism, however slight the risk of an attack with smallpox might be. Political considerations also played a role in the decision. Given the president's repeated warnings about bioterrorism, continued U.S. support for prompt destruction of the variola virus stocks stored at the CDC would have given conservative members of Congress an opening to accuse the administration of "unilateral disarmament" in the face of the terrorist threat.

Clinton's political instincts turned out to be correct. In early April, after he had made his decision but before the new policy had been announced publicly, the White House received a letter signed by seven leading Republican senators, including Senate Majority Leader Trent Lott. Dated March 22, 1999, the letter conveyed the senators' strong belief in the importance of retaining the variola virus stocks for biodefense research. "We understand that your administration will soon be deciding the fate of the remaining U.S. smallpox virus cultures held at the Centers for Disease Control and Prevention (CDC) in Atlanta," the letter began. "We believed back in 1995 that it was a bad idea to destroy the only *known* global stockpiles of the smallpox virus, when this plan was originally being urged by the World Health Organization (WHO). As the WHO now gears up to consider its postponed decision, we believe even more strongly that destruction of the U.S. smallpox cultures would undermine U.S. national security and would serve no public health purpose whatsoever. . . .

"You yourself, Mr. President, have made defending the United States against chemical and biological weapons a top national security priority. . . . A decision to destroy U.S. smallpox holdings would undercut your own initiatives to improve our preparedness in the face of these threats. . . . Mr. President, the national security, public health, and scientific case for retaining the existing and secure U.S. stockpile of smallpox virus cultures is even stronger and more compelling than five years ago. We urge you to reject the wholly hollow and symbolic destruction of smallpox virus which would only jeopardize the future health and security of the United States."

On April 22, 1999, the White House press office released an official statement announcing President Clinton's decision to delay indefinitely the destruction of the smallpox virus stocks stored at the CDC and noting that the decision had been based on a consensus recommendation of his advisors, reflecting agreement among all interested departments. The statement explained that the new policy "reflects our concern that we cannot be entirely certain that after we destroy the declared stocks in Atlanta and Koltsovo, we will eliminate all the smallpox virus in existence. While we fervently hope smallpox would never be used as a weapon, we have a responsibility to develop the drug and vaccine tools to deal with any future contingency—a research and development process that would necessarily require smallpox virus. In the end, we reached the conclusion we believe is most likely to reduce the possibility of future loss of life as a result of smallpox."

While conceding that the president's decision to retain the virus could harm relations with some allies and complicate U.S. foreign policy, a White House spokesman expressed hope that the decision would open up new possibilities for joint research with Russia. He declined to speculate about what would happen if a majority of WHO member states at the upcoming World Health Assembly in Geneva defied Washington and Moscow by voting to destroy the virus stocks. In fact, because the assembly had no power beyond persuasion to enforce its actions, it was unlikely to pass a resolution recommending immediate destruction if both the United States and Russia opposed the idea.

As a private citizen, D. A. Henderson spoke out strongly against the Clinton administration's decision. In a commentary written for the opinion page of the *Baltimore Sun,* he expressed his skepticism about the feasibility and value of developing anti-smallpox drugs and stressed the moral leadership that would accrue to the United States by supporting destruction of the virus stocks. "The steadfast U.S. position advocating destruction of the virus was not a capricious decision," the op-ed piece concluded. "Reversing that position now could be the ultimate folly." By this time, however, Henderson had become a voice in the wilderness. Ironically, his very success in raising the consciousness of policymakers about the threat of smallpox as a possible terrorist weapon had worked against his interest in seeing the virus stocks destroyed.

In May 1999, the fifty-second annual session of the World Health Assembly convened at the Palais des Nations, a sprawling limestone building in Art Deco style that had served as headquarters of the League of Nations during the 1930s. Peacocks strutted across the manicured grounds, which offered a view of Lake Geneva and—faintly visible through the haze—the jagged peaks of the Mont Blanc massif beyond. Security was tight because Kurdish terrorists had occupied the Palais des Nations a few months earlier. Metal barriers reinforced with coils of concertina wire blocked driveways leading to the building, and Swiss soldiers carrying automatic rifles manned security checkpoints and checked identity papers.

Ken Bernard was the White House point man at the Geneva conference. During his long career in the Public Health Service, he had served on several U.S. delegations to the World Health Assembly, but now he faced his most challenging assignment: persuading a skeptical world that the possibility of biowarfare and terrorism with smallpox warranted investing scarce health resources to develop new treatments for an eradicated disease. Before his departure for Geneva, some analysts in Washington had predicted that the developing world would rise up in fury against the United States and Russia for seeking to block the immediate destruction of the virus stocks.

Bernard made the rounds of the delegates' lounge at the Palais des Nations, conferring with various delegations and explaining the merits of the U.S. position. Most of the developing countries, particularly those that had suffered most from smallpox, made it clear that they wanted the WHO to destroy the virus stocks on schedule. Still, they listened with growing concern as he described the potential for a resurgence of smallpox from natural sources or deliberate use.

Bernard's strategy was to build a broad base of support for defensively oriented smallpox research by proposing to subsume it under an international program that would be closely monitored by the WHO. To this end, he organized an informal drafting group, chaired by a Swiss delegate, that prepared an alternative resolution authorizing the continued retention of the stocks of variola virus at the CDC and Vector for the purpose of developing antiviral agents and improved vaccines and conducting basic research. Drawing on a straw-man text that Bernard had composed, the draft resolution mandated the creation of an international scientific advisory group under WHO auspices that would approve and oversee smallpox research. All such activities would be funded outside the regular WHO budget through voluntary contributions by member states. Although the duration of the smallpox research program would be open-ended, the stated goal was to build an international consensus for the eventual destruction of the virus stocks.

Twenty-six countries agreed to be listed as cosponsors of the draft resolution, and a few others pledged to provide passive support by not objecting to it. The delegation of India, however, made clear that it was under strict instructions from New Delhi to press for immediate destruction of the variola virus stocks and hence could not accept the open-ended research program called for in the draft resolution. On the morning of May 21,

Bernard met for coffee in the delegates' lounge with two members of the Indian delegation and tried to work out a mutually acceptable formula. The Indians proposed amending the draft resolution so that smallpox research could continue only up to a specified deadline without another positive decision by the World Health Assembly. Bernard suggested a period of five years, but the Indians insisted on three. Sensing the potential for a breakdown in consensus in the international forum, Bernard recognized that the United States would have to compromise. India was a major leader of the G-77 group of developing countries, and without its support, the draft resolution might well go down to failure.

Bernard consulted with other members of the U.S. delegation. After some discussion, they supported his recommendation to accept the Indian demand limiting the smallpox research program to three years. Although the Pentagon representatives would have preferred more time, they agreed that it was important to demonstrate that the United States was acting in good faith to address its legitimate security concerns and did not have a hidden agenda. Now that a compromise acceptable to both the United States and India had been worked out, Bernard was fairly confident that the draft resolution, as amended, would be approved by the WHO member countries. Even so, the possibility of unanticipated objections remained.

On the afternoon of May 21, the technical committee reporting to the World Health Assembly took up the draft smallpox resolution. Early in the floor debate, the Indian representative, H. K. Singh, proposed an amendment authorizing "temporary retention up to but not later than 2002" of the existing stocks of variola virus at the CDC and Vector "for the purpose of further international research into antiviral agents and improved vaccines, and to permit high-priority investigations of the genetic structure and pathogenesis of smallpox." Over the next hour, delegates from twenty-five other countries spoke, most of them endorsing the draft resolution as amended by India. Although Iran, Zambia, and Nicaragua expressed strong reservations about delaying the destruction of the variola virus stocks, they did not attempt to block the resolution. By the end of the debate, it was apparent that a broad consensus had emerged, and the resolution was adopted by acclamation.

Although the resolution still had to be approved by the World Health Assembly in plenary session, that step was considered a mere formality: Variola virus had won another stay of execution. After the committee's decision, Ken Bernard was relieved and jubilant. Adoption of the draft resolution by consensus rather than by a politically divisive vote was the outcome he had most hoped for. Over the next two years, the World Health Assembly would review the progress of the authorized smallpox research. Then, in May 2002, the assembly would decide whether to give the variola virus stocks yet another reprieve or to move forward with destruction at the end of the year.

Staff members at the WHO Secretariat were disappointed that destruction of the variola virus stocks had been postponed yet again but pleased that the member countries had specified a three-year time frame for the smallpox research program. David Heymann, the director of the WHO's Division of Communicable and Emerging Diseases, tried to put the best face on the situation. "There is still a clause in [the resolution] that says it should be destroyed and we have no reason to think it won't be, at some future date," he said. Heymann, whose boyish face and full head of dark hair made him look at least ten years younger than his fifty-three years, had spent 1974 and 1975 working on the smallpox eradication program in India. Like other veterans of the campaign, he favored immediate destruction of the variola virus stocks. He also felt strongly that smallpox research should not divert scarce resources away from fighting ongoing epidemics such as AIDS, tuberculosis, and malaria, which were causing extensive mortality, suffering, and disability throughout the developing world.

In the aftermath of the World Health Assembly, Dr. Heymann began to implement the new resolution. His first task was to hire an internationally respected scientist who would be responsible for coordinating the small-pox research effort and organizing the WHO international scientific advisory committee that would approve and oversee specific research projects at the CDC and Vector. For this position, he recruited Dr. Riccardo Wittek, a well-known poxvirologist at the University of Lausanne, Switzerland. Because funds for smallpox research could not come from the WHO regular budget, the Swiss government agreed to pay Wittek's salary as a full-time consultant.

Although the WHO resolution called for international cooperation, the practical reality was that all smallpox research conducted at the CDC would have to be consistent with U.S. export control laws and regulations. Requests by non-U.S. scientists to conduct experiments with live variola virus would be considered on a case-by-case basis depending on past work and expertise, but only individuals with impeccable credentials would be granted hands-on access to the virus and anyone suspected of ulterior motives would be kept out. The Russians were equally reluctant to allow foreign scientists—with the exception of a few Americans—into the small-pox laboratory at Vector.

For Ken Bernard, the approval of the smallpox resolution by the World Health Assembly meant that it was time for the U.S. government to "fish or cut bait." Washington and Moscow had three years to show substantive progress toward developing new tools for fighting the disease. If little interest and investment in smallpox research materialized over the next several months, international pressure to destroy the virus stocks would probably become irresistible. It was also clear that because of the ongoing financial crisis in Russia, Vector would need U.S. government funding to conduct its share of the research program.

Kanatjan Alibekov (a.k.a. Ken Alibek) continued to warn that the former bioweapons scientists at Vector, such as Sandakhchiev, Shchelkunov, and Netesov, should not be trusted to work with live variola virus. Based on a

between-the-lines reading of the Russian scientific literature since his de-
fection in 1992, Alibekov had inferred that offensively oriented research at
Vector was continuing. Scientists have a strong incentive to publish to maintain their professional reputations, and the virologists at Vector were no
exception. Thus, during the 1990s, these individuals had disclosed some
of their unclassified results in Russian scientific journals. Although the papers made no direct reference to biological warfare, Alibekov found several indications that the Vector scientists were continuing to study the
molecular biology of variola virus with the aim of enhancing its military
potential. Examples of such "dual use" research included the creation of
a library of cloned fragments of variola DNA, from which infectious
smallpox viruses might be created in the laboratory; a complete analysis
of the smallpox genome to identify the functions of specific genes, including those responsible for virulence; and the genetic engineering of vaccinia and other poxviruses. In an op-ed piece for the *New York Times*,
Alibekov warned, "Without proper oversight and control, there is no way
to insure that smallpox research in Russia remains benign. American hopes
that binational cooperation will provide that oversight are a misreading of
history."

Alibekov also considered it "recklessly naïve" to believe that the two
official repositories in Atlanta and Koltsovo were the only ones extant and
that their destruction would eliminate variola virus from the planet. Because of these considerations, he did not favor destroying the known stocks
of the virus stored at the CDC and Vector. Instead, he proposed transferring them to a maximum-containment laboratory that would be built under
WHO auspices at a remote location in a neutral country, such as Switzerland. Scientists from several countries, including the United States and
Russia, would be selected to conduct research at this facility. The laboratory
would perform research on orthopoxviruses in general, with a particular focus
on smallpox, and would develop new drugs for the prophylaxis and treat-

ment of outbreaks caused by terrorist or military use. Because maximum-containment laboratories are extremely costly to build and maintain, however, it was unclear who would provide the money to finance such a facility. For that reason, the proposal seemed unlikely to be implemented.

While acknowledging Alibekov's concerns, U.S. government officials justified funding smallpox research at Vector as a means to benefit from the Russian scientists' depth of technical knowledge while keeping them gainfully employed, under close observation, and out of the clutches of would-be proliferators such as Iran. Beginning in 1995, the U.S. government had sought to counter the threat of "brain drain" to countries seeking biological weapons by providing grants for peaceful research to former Biopreparat scientists, including several at Vector. Such support was provided through two entities: the International Science and Technology Center in Moscow and the U.S. Civilian Research and Development Foundation. Despite Alibekov's warnings that offensive military objectives might still lurk behind Vector's research program, it was considered more important to enable the former weapons scientists to feed their families and live in dignity without being tempted to sell their deadly know-how to the highest bidder. As Army Colonel Dennis Duplantier, a Cajun from Louisiana, colorfully described the U.S. policy rationale: "If you want to milk a rattlesnake, you gotta grab it by the head."

From December 6 through 9, 1999, the new expert group mandated by the World Health Assembly resolution, the Advisory Committee on Variola Virus Research, met behind closed doors at WHO headquarters in Geneva. Carefully selected for expertise and geographical balance, the committee consisted of sixteen virologists from as many countries; D. A. Henderson was among them.

At the start of the meeting, scientists from USAMRIID, the CDC, and Vector presented their proposed topics for smallpox research, including the sequencing of additional variola strains, the screening of antiviral drugs, and the development of a safer vaccine. Henderson, supported by the committee members from India and Brazil, argued strongly that the research program should not be an open-ended quest for scientific knowledge, and that all projects should be designed for completion within the three-year time frame approved by the World Health Assembly. After considerable discussion over when the three-year period would actually come to an end, it was agreed that all research with live variola virus would have to be completed by midnight on December 31, 2002.

The WHO advisory committee also discussed the feasibility of developing a genetically engineered smallpox vaccine that would be safer to use in people with weakened immune systems. Several members objected that developing a next-generation vaccine could take as long as a decade, greatly exceeding the time frame of the research program. Moreover, since it would be impossible to test the vaccine in human subjects, one could never be confident that it would protect against smallpox until it was actually used. In view of these realities, the WHO advisory committee decided that developing a next-generation smallpox vaccine was not a realistic option. Although further work on a vaccine for at-risk populations should be encouraged, such research should not be dependent on access to live variola virus. Finally, the committee established a five-member "scientific subcommittee" to review specific smallpox research proposals and ensure their consistency with the agreed priorities and deadline. Because of the small size of this subcommittee, it would be able to turn around proposals within one month of receipt.

Despite the green light from the WHO advisory committee, it took several months for the U.S. smallpox research program to get under way. One reason for the delay was institutional resistance from the CDC, where all work with the live virus would have to take place. Early proposals to do some of the research at USAMRIID had been rejected for several reasons: the safety and security problems associated with transferring samples of live variola virus from the CDC in Atlanta to Fort Detrick in Maryland, the negative political ramifications of conducting smallpox research at an army laboratory, and concern about setting a bad example for the Russians.

Dr. James M. Hughes, the director of the CDC's National Center for Infectious Diseases (NCID), was reluctant to devote scarce space and time in the center's maximum-containment laboratories for research on smallpox when emerging infections posed a far more immediate threat to public health. Both of the "hot labs" at the NCID were in constant demand for studies of deadly, incurable viruses such as Ebola hemorrhagic fever, which continued to claim victims in developing countries. It was also impossible to predict when the next outbreak of a new infectious disease somewhere in the world would require the urgent characterization of the causative agent. In 1993, for example, a mysterious and often fatal type of pneumonia called "Sin Nombre" had appeared in New Mexico, Arizona, and Colorado. Within days, the CDC had identified the source as a mouse-borne variant of hantavirus that had never before been seen in the United States. CDC scientists had also conducted an emergency outbreak investigation in 1999, when the previously unknown Nipah virus had infected pig populations in Malaysia and then spread to humans, causing more than one hundred deaths. If one of the hot labs at the CDC was set aside for smallpox research, other important tasks would have to go begging. Hughes liked to say that the Pentagon had the resources to fight two major wars at the same time, but the CDC did not.

Ken Bernard, irritated that Hughes and other CDC officials were dragging their feet on smallpox research, made it clear that the program was a top priority of the Clinton administration. Although the CDC might have a tradition of independent inquiry, it was still a U.S. government laboratory and was duty-bound to follow the policy guidance set at the highest levels. Hughes finally got the message when a congressional committee identified a misappropriation of funds at the NCID that was largely attributable to sloppy accounting practices. This minor financial scandal, together with the letter to President Clinton from seven Republican senators expressing strong support for smallpox research, brought home the political reality that unless the NCID got into line it could be vulnerable to further congressional inquiry. Having read the tea leaves, Hughes declared his support for a "legitimate research agenda" on smallpox, including an initial three-month block of time in one of the maximum-containment laboratories.

In April 2000, nearly a year after the World Health Assembly decision, two army scientists from Fort Detrick began working with live variola virus at the CDC. Peter Jahrling was trying to develop an animal model of smallpox in monkeys, while John Huggins was screening another set of antiviral drugs for their effectiveness against variola. The clock was ticking: Only two and a half years remained to complete all of the planned experiments with the live virus before the WHO deadline of December 31, 2002.

The decision to conduct all research with live variola virus at the CDC meant that Jahrling and Huggins had to spend three-month blocks of time in Atlanta, six hundred miles from their families in Maryland. Although the two army scientists found it stressful and inconvenient to be away from home for such an extended period, they felt that they had been given a precious window of opportunity to make some progress in limiting the smallpox threat. By working fourteen-hour days, seven days a week, they managed to get a great deal accomplished.

The research took place in Building 15, a tall, modernistic red brick structure at the heart of the CDC campus that contained enough reinforced concrete to withstand an earthquake, a tornado, or an airplane impact. Access to the facility was tightly controlled and the interior rooms were continually monitored with closed-circuit TV cameras. In addition to several high-containment ("Biosafety Level 3") laboratories, Building 15 contained two maximum-containment ("Biosafety Level 4") suites for working with lethal viruses for which no vaccine or treatment was available. Although a vaccine against smallpox existed, extreme precautions were now required when working with variola virus because of the civilian population's near-total lack of immunity.

Each hot lab was a totally artificial environment of bright lights, stainless steel, and white ceramic tile, akin to the interior of a superclean submarine or a futuristic space station. Multiple redundant safety measures served to prevent deadly microbes from escaping into the surrounding urban environment. All exposed surfaces were bathed in ultraviolet light, which killed viruses, and the ventilation system was designed so that the atmospheric pressure decreased toward the center of the laboratory, causing air currents to flow inward and then up and out through a set of high-efficiency particulate air (HEPA) filters in the ceiling. Thus, if a breach in containment were to occur, the virus-contaminated air would be retained inside the lab and not allowed to escape into metropolitan Atlanta.

In an anteroom outside the hot lab, scientists and technicians removed their street clothes, donned surgical gowns and gloves, and climbed into blue plastic "space suits" that isolated them completely from the deadly agents with which they worked. Each suit had bright yellow rubber gloves securely taped at the wrists, a clear visor hood, and an air filter with a supply valve protruding from the side. All workers entered and departed the laboratory through an air lock. Inside the hot lab, coiled red plastic air hoses hung down from the ceiling every few yards. The space-suited scientists moved from station to station by unhooking from one hose and attaching to the next to keep their suits inflated with air. On leaving the lab, each worker returned to the air lock and stood for four minutes under a chemical shower, allowing the green disinfectant spray to drench the outside of the plastic suit and kill any invisible virus particles clinging to it.

The floors above and below the maximum-containment suites were devoted to maintaining the integrity of the biocontainment system at all times. On the floor above, banks of ventilator fans equipped with HEPA filters screened out microscopic particles from the air flowing in and out of the laboratories. The floor below housed a diesel backup generator, rows of backup air-supply tanks, and a pair of thousand-gallon tanks to sterilize all liquid waste from the laboratory at high temperature. After the waste had cooled, it was released into the Atlanta sewer system.

The study that Peter Jahrling performed in the hot lab sought to determine whether monkeypox in cynomolgus macaques, which caused a pustular rash and was often lethal, was a good model of human smallpox. Earlier, at the Armed Forces Institute of Pathology in Maryland, he had examined tissue specimens obtained at autopsy from smallpox victims and preserved in blocks of paraffin. In these specimens, variola virus was most concentrated in organs rich in macrophages (a type of scavenging white blood cell), including the spleen, the lymph nodes, the thyroid, and the Peyer's patches (clumps of lymphoid tissue associated with the intestines). Because the pathological effects of smallpox in humans were not identical to the manifestations of monkeypox in monkeys, this finding raised the question of whether monkeypox offered a realistic model of the human disease. The only way to find out was to attempt to induce smallpox in monkeys and then compare its pathological effects with those of monkeypox.

The scientific literature indicated that although monkeys could be infected by variola virus, which was detectable in their blood, they did not develop a fever or skin rash similar to the symptoms of smallpox in humans. Nevertheless, Jahrling hoped that when monkeys were exposed to a high-dose aerosol of a virulent strain of variola major—the Yamata strain, which Fort Detrick had studied as a potential biological weapon during the 1960s—they would develop a disease similar to human smallpox. If smallpox and monkeypox then turned out to have similar pathological effects in monkeys, the FDA might accept monkeypox as a model for testing antiviral drugs and vaccines. The initial results of the experiment were disappointing: The macaques became infected with variola virus but did not develop a severe or fatal illness resembling human smallpox. In June 2001, however, Jahrling succeeded in inducing fever and a characteristic smallpox rash in monkeys by simultaneously exposing them to a concentrated aerosol of variola virus and injecting the virus into the animals' bloodstream. The following month, he obtained identical results by intravenous injection alone. Twelve of the fourteen animals in the second experiment died of the disease, whose pathological effects closely resembled those of clinical smallpox. This breakthrough suggested that an animal model of the human disease might be possible after all, although Henderson remained skeptical because of the unnatural route of infection.

Meanwhile, John Huggins continued to screen candidate anti-smallpox drugs in cell culture by measuring their ability to inhibit the replication of several strains of variola virus. The most promising compounds would then be tested in two animal models: mice infected with cowpox, followed by monkeys infected with monkeypox. Huggins also planned to study the effects of the candidate drugs against several other poxviruses, including vaccinia and mousepox, in the hope of finding a reliable surrogate that could take the place of live variola virus in future screening experiments.

The smallpox research program at Vector did not get started until March 1, 2001, nearly two years after it had been authorized by the World Health Assembly in May 1999, because of a series of bureaucratic hurdles. First, developing and refining the research proposals required several meetings between U.S. and Russian officials in Washington and Koltsovo. It then took six months for the Russian Ministry of Health to approve the research plan, which was reviewed in September 2000 by the WHO Advisory Committee on Variola Virus Research. Once the WHO panel had signed off on the proposal, it was necessary to develop equipment lists and technical work programs and submit them for final approval by the U.S. Interagency Working Group.

The U.S. government budgeted a total of $5 million for the smallpox research program at Vector, of which $2 million was appropriated in fiscal year 2000, split equally between the departments of HHS and Defense. As of March 2001, three projects had been approved and two bilateral agreements signed. All of the U.S.-Russian project agreements stipulated that financial support for work with live variola virus would end by the WHO-imposed deadline of December 31, 2002. If the World Health Assembly decided in May 2002 to extend the period of smallpox research, the contracts would be renegotiated. David Heymann at the WHO was impatient for the work at Vector to get under way as soon as possible, so there would be no excuse for retaining the variola virus stocks beyond the end of 2002.

The first signed agreement between the United States and Russia covered a joint effort by Vector and the CDC to determine the DNA sequence of additional strains of variola virus, thereby facilitating the development of DNA "probes" for more precise diagnosis and detection. From its collection of 120 isolates, Vector planned to select for partial sequencing some forty strains obtained from smallpox outbreaks of varying severity that had occurred in different years and geographical locations. The Russian scientists would cultivate each strain in cell culture, extract the viral DNA and cut it into snippets, and clone the snippets in bacteria. Some of the cloned smallpox DNA would be sent to the CDC, which had advanced sequencing machines. For each strain of variola, the plan was to sequence initially only a few segments near the terminal regions of the genome. The small fraction of variola strains that differed significantly from the others would then be sequenced in their entirety. Vector and CDC researchers also planned to identify viral genes that were potential targets for anti-smallpox drugs and to sequence these genes across a range of isolates.

The second project at Vector involved screening hundreds of candidate antiviral drugs, initially in cell culture and then in suckling mice less than twelve days old, which could be infected with variola. Two additional projects were planned: the production of "combinatorial libraries" of antibodies specific to variola for diagnostic purposes, and the development of a "gene chip" to diagnose various strains of smallpox virus, as well as other orthopoxviruses and herpesviruses.

The Russian government turned down one collaborative research project suggested by the U.S. side. Peter Jahrling of USAMRIID proposed a joint effort to develop a non-human primate model for smallpox in hamadryas baboons, which do not die of variola infection but develop fever and significant illness. Because the maximum-containment laboratory at the CDC was too small for research with such large primates, the proposed study would have been an ideal use of the far more spacious facility at Vector. But Moscow rejected Jahrling's proposal without providing a clear rationale. Some U.S. scientists speculated that Vector had already developed an animal model for smallpox in baboons during the Soviet era and did not wish to share its unique knowledge in this area.

Although the U.S. government tried to recruit American scientists to conduct collaborative smallpox research at Vector, the small community of poxvirologists at the CDC and the National Institutes of Health expressed little interest. Dr. Michael Bray, who worked with John Huggins at USAMRIID, planned to spend some time in Koltsovo working on the drug-screening studies; in December 2000, he did his initial training in the use of a Russian isolation suit. But Huggins and the CDC's Joe Esposito had no desire to spend a few months working at Vector.

U.S. government funding for Russian smallpox research came with a congressionally imposed requirement for external oversight to make sure the money was being spent as intended. Although the United States would not have a full-time scientist stationed in Koltsovo, HHS assigned Alex Kosyak, a pharmacist and commissioned officer in the U.S. Public Health Service who spoke fluent Russian, to oversee and facilitate the technical work at Vector for the duration of the smallpox research program. HHS also planned to install computer terminals with high-speed Internet connections inside the maximum-containment laboratory in Building 6, enabling the researchers to take notes and send e-mail messages. The Internet connection would support a digital video camera for continuous observation and monitoring of the smallpox work. Because the maximum-containment laboratory did not have windows, the video monitoring system would make it possible to respond promptly in case of an accident or other emergency.

By the time the smallpox work at Vector got under way, little time remained to complete the ambitious research agenda before the WHO's December 2002 deadline to halt work with live variola virus. The Russian laboratory shut down in the summer because of the lack of air conditioning, at which time scheduled maintenance was performed and most of the staff went on vacation. During the winter, the laboratory sometimes ran short of steam and other utilities and was forced to halt operations. Given these time constraints, U.S. and Russian officials began to worry that the essential studies with the live virus might not be completed by the WHO-imposed deadline.

Lev Sandakhchiev, for his part, hoped that the destruction of the variola virus stocks would be postponed after 2002, ideally forever. He strongly believed that destroying the virus would be "a great scientific and political mistake," both because of the need to defend against the threat of bioterrorism and because he considered it "inevitable" that some other orthopoxvirus, such as monkeypox, would eventually mutate to fill the ecological niche vacated by the eradication of smallpox. Among other projects, Sandakhchiev advocated a long-term research effort to develop a genetically engineered vaccine that would be safer in HIV-infected people and others with impaired immune function. One possibility would be to devise a multifunctional vaccine that protected against hepatitis B and C, as well as smallpox. In that case, it would be politically feasible to administer the combined vaccine in advance of a smallpox outbreak because hepatitis posed a more immediate risk of infection.

D. A. Henderson and other members of the WHO Advisory Committee on Variola Virus Research remained adamantly opposed to extending the deadline for research with the live virus. Henderson noted that although the primary justification for retaining the variola virus stocks was to de-velop an antiviral drug and a safer vaccine, their utility for dealing with a public health emergency was far from evident. It also appeared likely that if the United States and Russia asked the World Health Assembly in May 2002 to extend the research deadline to permit the completion of ongo-ing studies with live variola virus, they would face strong political resistance from India, Brazil, and other developing countries. Margaret Hamburg be-lieved that the final outcome would depend on how promising the small-pox research agenda proved to be and how rapidly it moved forward. "My guess, given the difficulty of working with the virus," she observed, "is that May 2002 will come around before people feel they've had a chance to ex-plore all of the scientific issues, and there will be another heated debate."

TWELVE

THE UNFINISHED CONQUEST

The WHO smallpox eradication campaign was a landmark achievement of international health. It vanquished an ancient scourge, ended the risky process of routine vaccination, and saved $2 billion dollars a year on the costs of prevention. Yet, in a bitter irony, smallpox eradication and the resulting decision to halt the vaccination of civilians worldwide had the unintended effect of enabling the Soviet Union—and, perhaps, other countries—to transform variola virus into an instrument of biological warfare, as it had been during the eighteenth century before Jenner's vaccine became widely available.

The bioterrorist threat from smallpox should not be exaggerated: "Rogue" states or terrorists seeking to acquire and use the virus as a weapon would have to overcome significant obstacles. Because smallpox has been eradicated from the wild for more than two decades and laboratory reconstitution of variola DNA into an infectious virus remains beyond the state of the art, the international black market would offer the only means of obtaining samples of the virus. Moreover, a government that retained or acquired smallpox would probably be deterred from using it in an overt military attack by the prospect of devastating retaliation—although a national leader who faced total defeat in war might consider unleashing the virus in a final act of revenge. Smallpox would also be an unlikely weapon for small, unsophisticated terrorist organizations. Only a state-sponsored terrorist group that was confident of concealing its identity, or a wealthy apocalyptic cult such as Aum Shinrikyo, would probably have both the motivation and the resources to launch a successful attack with the virus.

Assuming that terrorists were able to gain access to a virulent strain of variola virus, they would have to surmount a series of technical hurdles: obtaining enough smallpox vaccine to protect themselves from infection, growing variola virus in eggs or tissue culture, and dispersing it effectively. During the 1960s, when the United States still had an offensive biowarfare program, the U.S. Army conducted secret field experiments to assess the nation's vulnerability to a clandestine smallpox attack. In an exercise in May 1965 at Washington's National Airport, army experts fitted briefcases with tiny aerosol generators that sprayed a harmless simulant with physical properties similar to those of variola virus. The results of this test suggested that one in every twelve travelers passing through the airport would have become infected with smallpox, rapidly dispersing the contagious disease to cities around the country.

Given the significant technical challenges involved in acquiring and delivering variola virus, the risk of a deliberate reintroduction of smallpox remains quite low, but it is not zero—particularly if terrorists received assistance from a former Soviet scientist who had done military work on the virus at Vector or Zagorsk. Moreover, despite the low probability of a deliberate release of variola virus, the consequences of such an attack could be catastrophic.

In February 1999, the Center for Civilian Biodefense Studies, a think tank that D. A. Henderson had founded the previous year at Johns Hopkins University, organized a national symposium on bioterrorism. Held in

Arlington, Virginia, this meeting was attended by some 950 public health experts, medical practitioners, and government officials from throughout the United States and abroad. The second day of the symposium featured a lengthy role-playing exercise based on a scenario in which an imaginary terrorist group covertly releases an aerosol of variola virus in a medium-sized American city. At four points during the simulation, representing the elapse of a few weeks to several months after the initial attack, Dr. Tara O'Toole, the principal author of the scenario, read portions of the narrative aloud. Four panels of hospital administrators and representatives from city, state, and federal government agencies then tried to respond to the unfolding crisis as if it were actually happening and they were trying to contain the epidemic. The aim of this role-playing exercise was to explore the diverse and often unexpected problems that might arise during a bioterrorist incident involving smallpox.

According to O'Toole's fictional but realistic scenario, on April 1, the U.S. vice president visits Northeast, a city of 2.5 million people, to give a major speech at the local university. A large crowd gathers in the university auditorium and hundreds more wait outside, where the vice president stops to shake hands and respond to questions from the media. The U.S. Secret Service and the Federal Bureau of Investigation (FBI) have information suggesting a possible threat against the vice president from a terrorist organization with suspected links to a rogue state. Members of this group are known to have made inquiries in the past about acquiring biological agents, including smallpox, and are suspected of having procured an aerosolization device small enough to conceal in the trunk of a car. The federal agencies decide, however, that the information—from a highly classified source—is too sensitive to share with the Department of Health and Human Services, the state health department, or local law enforcement. Indeed, state and local government officials have never been consumers of classified intelligence information, nor have security and clearance procedures been established for that purpose.

On April 12, eleven days after the vice president's visit to Northeast, a twenty-year-old student arrives at the university hospital emergency room with a fever of 103 degrees Fahrenheit and severe muscle aches. She is pale and has a slightly lowered white blood cell count, but her physical exam and laboratory results are otherwise normal. Diagnosed with a presumptive viral infection, she is sent home with instructions to drink fluids and take an analgesic for muscle aches. Later that day, a forty-year-old electrician arrives at the emergency room with severe lower backache, headache, shaking chills, vomiting, and a temperature of 102 degrees F. Again, the patient is discharged with an analgesic and instructions to drink fluids. The following morning, four young adults in their twenties come to the university hospital emergency room with flulike symptoms and are likewise sent home.

On April 14, the female student returns to the emergency room after collapsing in class. She has developed a rash of blisters on her face and arms and is admitted to an isolation room with a presumptive diagnosis of adult chickenpox. The next day, the electrician returns to the emergency room by ambulance suffering from a skin rash, high fever, and prostration. He, too, is admitted to an isolation room with a presumptive diagnosis of chickenpox. Smallpox is not suspected because none of the clinicians at the hospital has ever seen an actual case.

At 6:00 P.M. that evening, an infectious disease specialist examines the two patients, both of whom have developed a marked rash on their face, arms, hands, and feet. The skin blisters are evolving into pustules, raising the possibility of smallpox. After taking swab specimens of the skin lesions, the doctor sends them to the laboratory for examination by electron microscopy. By 7:00 P.M., the lab reports the presence of an orthopoxvirus consistent with variola. Fifteen minutes later, the hospital epidemiologist declares a contagious disease emergency. Both patients are moved to special isolation rooms equipped with negative-pressure ventilation systems and high-efficiency particulate air filters, and visitors and other members of the hospital staff are forbidden to enter the floor. Infection-control nurses begin interviewing the staff to determine who has been in face-to-face contact with the two patients during the initial emergency room visits and admission.

At 7:45 P.M., the chair of the department of medicine and the hospital president meet with the infectious disease physician and other members of the hospital staff, including the epidemiologist, the vice president for public relations, and the general counsel. The city and state health commissioners join the meeting by phone and discuss the need to vaccinate and isolate all possible contacts of the two patients in the hospital. After the conference call, the state health commissioner calls the CDC to request that smallpox vaccine be released for the hospital staff and patient contacts. Because vaccine supplies are limited, the CDC insists that the diagnosis of smallpox first be confirmed. To this end, specimens from the two patients are flown by military aircraft to Atlanta for analysis in the CDC's maximum-containment laboratory.

Because acts of domestic terrorism are under the jurisdiction of the federal government, the state health commissioner notifies the FBI. The bureau requests that the city police be called in to help maintain order and to make sure that no patients, staff, or visitors be allowed to leave the hospital until they have been identified and their addresses recorded, so that they can be vaccinated as soon as the smallpox vaccine has been released by the CDC. Hospital security and local police seal the doors of the hospital and redirect ambulances to other medical centers. Visitors and staff are confused and angered by the refusal of the police to allow anyone to leave, especially when no clear explanation is given. Rumors spread that a deadly contagious disease has broken out in the hospital, or that an armed terrorist is at large. As nerves begin to fray, a scuffle breaks out between police and visitors desperately trying to leave the building; three people are injured and sent to the emergency room.

More police and FBI agents arrive and surround the hospital, and local television vans arrive to report the breaking story for the late-night news.

The hospital public relations officer explains to journalists that the lock-in is only temporary and designed to gather names and addresses so that people can be notified and treated if cases of a suspected, but unnamed, contagious disease are confirmed. Wild rumors continue to circulate about what the unnamed disease might be, with Ebola being the favored candidate.

At 11:30 P.M., the patient specimens arrive at the CDC in Atlanta; by midnight, the diagnosis of smallpox has been confirmed. At that point, a conference call is set up involving the university hospital staff, the city police chief, the state health commissioner, the state attorney general, the governor, the CDC, the FBI, an assistant secretary from HHS, and staff members from the National Security Council and the White House. Much of the discussion concerns how to release the diagnosis to the news media. Finally, the mayor and the governor agree to go on television the next morning with the health commissioner; the FBI director will also make a statement, and the President will address the country at noon. These messages must be carefully crafted to avoid inciting panic: If the initial flu-like symptoms of smallpox are described on the television news, hordes of noninfected but anxious people with colds or headaches could swamp emergency rooms across the state. Meanwhile, the CDC makes arrangements to release supplies of smallpox vaccine early the next morning to vaccinate patient contacts and health practitioners caring for the two hospitalized victims.

On the morning of April 15, a conference call is set up among the CDC, the FBI, HHS, the National Security Council, and state health authorities. Federal officials now believe that a covert bioterrorist attack involving smallpox occurred two weeks earlier in Northeast. They fear that other releases of aerosolized virus might have taken place but have not yet come to light, and that further attacks might be imminent. A complete quarantine of Northeast is considered but rejected as neither feasible nor warranted. Meanwhile, the CDC is reluctant to begin a mass vaccination campaign until the dimensions of the outbreak are better understood. The CDC agrees to send three Epidemic Intelligence Service officers to assist the state epidemiologist in assessing the outbreak. In the meantime, the authorities decide to vaccinate the members of the hospital staff, the visitors to the floor where the two patients were admitted initially, and all face-to-face patient contacts, such as family members and coworkers. City health officials will compile a registry of these contacts and monitor them daily for fever. After further discussion, the decision is made to vaccinate all health care personnel, first responders, and police in Northeast and any other cities in the state that report confirmed cases of smallpox.

At noon on April 15, the president goes on television to inform the nation that a bioterrorist attack has been carried out by unknown perpetrators. He vows that the assailants will be identified and brought to justice, and urges calm and cooperation with government authorities. City and state public health officials will attempt to identify and vaccinate everyone who attended the vice president's speech in Northeast. Reporters point out, however, that the government has no idea how many people are sick or how widespread the epidemic might be.

Meanwhile, university hospital staff search emergency room records and attempt to contact all patients with fever who were seen during the previous week. This method turns up three more likely cases of smallpox; telephone follow-up reveals that one of them has been admitted to another hospital out of state. The state health department also activates a preexisting phone tree to query all hospitals and walk-in clinics throughout the state about smallpoxlike cases. This active surveillance effort identifies an additional eight patients admitted with fever and blister rash. All of them are extremely ill and two are delirious.

CDC and state health officials discuss possible quarantine strategies for managing the epidemic if the amount of vaccine available is not sufficient to cover all patient contacts, as seems likely. Mandatory isolation of unvaccinated contacts at home is considered, but the legal basis for this policy is questionable. Instead, the authorities decide to convert the university hospital in Northeast into an isolation facility that will accept transfers of small-pox patients hospitalized at other locations around the state.

By the evening of April 16, thirty-five more cases have been identified in eight emergency rooms and clinics around the city, and ten cases have been reported in an adjoining state. The CDC alerts all state health departments to be on alert for possible smallpox and urges prompt and strict isolation measures. States are instructed to send specimens from suspect patients to CDC headquarters in Atlanta for confirmation.

On April 17, the city and state health departments vaccinate ten thousand residents of Northeast, with assistance from volunteer physicians and nurses. Because of evidence suggesting that immunization within three days of exposure provides at least partial protection against smallpox, the containment strategy calls for vaccinating all individuals who have been in close contact (within six feet) of a patient or who are expected to be within this range, such as emergency medical teams and medical care staff who are seeing sick patients. Before vaccinating contacts, public health officials weigh the likelihood that a person has been exposed to smallpox against the potential for adverse reactions to the vaccine, and attempt to make an informed medical judgment on whether or not to proceed.

The following day, an additional twenty thousand residents of Northeast are vaccinated. State health officials continue to press for a statewide vaccination campaign, and unions representing nurses and other health care workers demand vaccine for all employees whose jobs involve direct patient contact. On April 19, the student with the first diagnosed case of smallpox dies. Meanwhile, ten more patients have been identified, bringing the total number to almost seventy. Confirmed cases have been identified in four states, with suspected cases in five more. On April 20, at the close of the fourth day of the vaccination campaign in Northeast, a total of eighty thousand people have been vaccinated.

By April 29, the CDC receives reports of one hundred new cases of possible smallpox, sixty in the original state and the others scattered over eight more states. Two cases in Montreal and one in London are also reported. The CDC and health agencies realize that they are seeing a second wave of smallpox cases. Although the latest victims were presumably infected by contact with people who attended the vice president's speech in Northeast, a second bioterrorist attack cannot be ruled out. Despite efforts to coordinate the criminal and epidemiological investigations, the perpetrators of the original attack have not been identified or apprehended.

Having received thousands of requests for smallpox vaccine from individual physicians, the CDC announces that the vaccine will be distributed only through state health departments. Because of the limited supply, priority will be given to vaccinating those at greatest risk of infection, such health care workers. Federal officials also announce a crash program designed to reduce the time needed to manufacture and approve more smallpox vaccine from three to two years. As the epidemic spreads, friction increases among the various levels of government. The governors of a dozen states call the White House demanding vaccine, and a state attorney general announces a suit against the federal government to force the release of vaccine for his state.

By May 1, more than seven hundred cases of smallpox have been reported. The CDC announces that 90 percent of the remaining U.S. vaccine stocks will be distributed to the affected states, but cautions that the available supply can cover only 15 percent of those states' populations. Governors are to establish state-specific priorities and mechanisms for vaccine distribution. By now, however, the number of potential contacts is so great that there is no longer any practical alternative to making the vaccine widely available in infected areas. Health officials also lack the time or resources to target vaccination precisely to those at greatest risk. As a result, the demand for vaccine rapidly exceeds the available supply.

Media reports begin to criticize the government's handling of the crisis, and dramatic television footage of children covered with pustules drives thousands of panicky people to emergency rooms, HMOs, and doctors' offices to demand vaccination and the evaluation of fever and other symptoms. This influx of patients, most of whom are "worried well," impedes the ability of state health departments and the CDC to determine the number of new cases. The health authorities decide to issue a recommendation that feverish patients who cannot be definitively diagnosed should be isolated and observed until their fever subsides.

From May 15 to 29, the third wave of the epidemic occurs. Over this two-week period, seven thousand more cases are reported in Northeast, several other U.S. states, and two foreign countries. Because extensive vaccination campaigns are under way, the nation's stocks of vaccine are rapidly dwindling. Some physicians attempt to stretch the remaining supply by diluting the vaccine or by employing the eighteenth-century technique of arm-to-arm vaccination, which involves taking material from the vaccination pustule on the arm of one individual to vaccinate another. Finally, however, the nation's stocks of smallpox vaccine are exhausted. From now on, the only available control method is isolation, which slows but cannot stop the spread of the disease.

In Northeast, the epicenter of the epidemic, the number of confirmed smallpox patients requiring isolation has long since outstripped the capacity of local hospitals to accommodate them. Smallpox cases and suspected contacts are now being isolated in the local armory and convention center, where volunteer physicians and nurses are providing palliative treatment. These health care providers are physically and emotionally exhausted from the long hours of work and seeing a third of their patients die. The epidemic has also had a severe economic impact on the city. Numerous business conventions scheduled months in advance have been canceled and the tourist trade is at a standstill. Attendance at theaters and sports events has collapsed, and many small businesses downtown are failing because suppliers and customers refuse to visit the area.

The president declares those states with the largest numbers of smallpox victims to be disaster areas, releasing federal funds to cover some of the costs of the epidemic. With the nation's supply of vaccine exhausted, Americans experience a pervasive sense of dread and foreboding. Most of the affected states close their public schools early because parents are keeping their children at home and teachers are refusing to come to work. Across the country, people fail to show up for jury duty or to attend public meetings for fear of contracting smallpox. In hospitals and HMOs where the staff have not been vaccinated, health care personnel have staged protests and some have walked off the job.

On May 30, the fourth wave of cases begins, and by mid-June, a total of fifteen thousand cases of smallpox have been reported in twenty U.S. states and four foreign countries. More than two thousand people have died, including two members of the vice president's staff and a Secret Service agent. The city of Northeast, hardest hit by the epidemic, has experienced several outbreaks of civil unrest. National Guard troops have been called in to help the police maintain order and to guard the facilities where smallpox cases and contacts are quarantined. Meanwhile, the mayor has been hospitalized with a heart attack.

As the epidemic continues, rumors of miracle treatments, fueled by the media, lead to cycles of unrealistic hope followed by bitter disappointment. Because it will take at least two years before the large-scale production of smallpox vaccine can begin, a few beleaguered physicians resort to the ancient technique of variolation—the preventive inoculation of healthy people with pus from smallpox lesions.

Meanwhile, the disease has been reintroduced into several regions of the world. Lacking adequate supplies of vaccine, foreign governments institute strict isolation procedures and enforce them harshly: Human rights organizations report numerous cases of smallpox patients being abandoned to die, and of recovering patients being denied housing and food. In response to rising civil unrest, a few countries impose martial law. Domestic and international travel fall off sharply and tourists shun smallpox-infected areas. Some countries refuse to admit U.S. citizens without proof of a recent smallpox vaccination, while others impose fourteen-day quarantines on all persons entering the country from abroad. A lucrative black market in falsified vaccination certificates develops. By year's end, endemic small-pox has been reestablished in fifteen countries, and the WHO director-general schedules a debate for the next World Health Assembly on whether to relaunch the global smallpox eradication campaign.

As this grim scenario played itself out, the role-playing participants became caught up in the tension and drama of the imaginary smallpox epidemic. Despite their best efforts to contain the outbreak, the disease managed to escape their control: Secondary transmission got out of hand, the available supply of vaccine was too limited to contain the spread of the disease, and the standard response plans did not work. The failure of con-tainment meant that the incident escalated rapidly from a local outbreak to a national epidemic and finally to a global pandemic. "We blew it," ad-mitted Dr. Michael Ascher, a California Department of Health official who participated in the role-playing exercise. "It clearly got out of control. Whatever planning we had . . . it didn't work. I think this is the harsh real-ity, what would happen." Although the Johns Hopkins scenario was inevi-tably speculative, it made clear the importance of stockpiling adequate amounts of smallpox vaccine and of maintaining the industrial capacity to produce more on short notice.

As of early 1999, the U.S. supply of smallpox vaccine was inadequate to cope with even a medium-sized outbreak that might result from a bioterrorist attack. The entire stockpile consisted of 150,000 vials of freeze-dried vaccine, which Wyeth-Ayerst Laboratories had manufactured during the 1970s under the trade name Dryvax and now kept in cold storage at a secluded plant near Marietta, a small town in Lancaster County, Pennsylvania. In an un-marked building at the plant site, behind several locked doors, a walk-in freezer was stacked to the ceiling with wooden crates, each filled with card-board boxes containing forty-eight small glass vials of freeze-dried vaccine.

As required by the FDA, Wyeth tested a few vials from the stockpile once a year to ensure that they remained potent. A third of the vials had developed some condensation, indicating that the vacuum seal had been breached, but testing showed that the vaccine still retained its immunizing power after twenty or thirty years and the affected vials were not discarded.

During the smallpox eradication campaign, each vial of Dryvax had yielded a maximum of one hundred doses when delivered by an expert vaccinator with a bifurcated needle. With less skilled vaccinators, however, the number of doses per vial was usually between fifty and seventy-five. This observation led to the conservative estimate that the stockpile of 150,000 vials stored in Marietta contained roughly 7.5 million doses—far from adequate for a U.S. population of 280 million people. Worldwide, supplies of smallpox vaccine were also limited. After the certification of smallpox eradication in 1980, the WHO had continued to store 200 million doses of vaccine in rented storage lockers at two locations in Switzerland and had regularly tested the stockpile for potency. The costs of vaccine storage and testing were significant, however, and the WHO budget was under stress at the time because of a failure by the United States to pay its dues. Thus, in 1990, twelve years after the last known case, the WHO had decided to reduce its stockpile of smallpox vaccine to 500,000 doses as a cost-saving measure, and the balance was sent back to the various donor countries.

Individual countries now reported possessing some 80 million doses of vaccine, but because of inadequate storage and testing, only about 50 million doses were believed to be of acceptable potency. South Africa had the largest stockpile, about 38 million doses, and Russia also had sizable stocks, albeit of uncertain quality. India had 5 million doses that had spent at least one summer without refrigeration. In any event, no country possessed enough vaccine to protect its own population from a major outbreak, let alone to provide assistance to others.

In the United States, the only effort to acquire additional smallpox vaccine had been made by the Department of Defense. In June 1997, the U.S.

Army had let a contract for 300,000 doses of freeze-dried smallpox vaccine under the Joint Vaccine Acquisition Program (JVAP). Given the contagious nature of smallpox, the small size of the military order was hard to explain. Reportedly, the Pentagon had based its requirement on the need to protect a limited number of troops being deployed into a war zone where they might face enemy use of the virus.

When none of the major U.S. pharmaceutical companies showed interest in manufacturing the amount of vaccine sought by JVAP, the Pentagon awarded the prime contract to a British-American joint venture called DynPort, based in Frederick, Maryland. Because the traditional method of growing vaccinia virus in the skin of living calves no longer met FDA quality standards, the new vaccine would have to be produced with modern cell culture techniques. DynPort subcontracted the manufacture of the army vaccine to another company called BioReliance Corp., headquartered in Rockville, Maryland. Altogether, process development, FDA licensure, and production were expected to take until 2006. Moreover, although the Wyeth vaccine had cost a penny per dose back in 1975, DynPort's quoted price was a stunning $77 per dose.

In mid-1999, in response to the growing concerns over bioterrorism, the Department of Health and Human Services decided to procure its own reserve stockpile of smallpox vaccine for immunizing large numbers of civilians in the event of a terrorist attack. Based on a computer simulation of a major release of variola virus in a U.S. city and the ensuing spread, CDC epidemiologists estimated that the civilian vaccine stockpile should contain 40 million doses, backed up with a "warm" production line capable of manufacturing at least 60 million additional doses as required.

In late August 1999, CDC Director Jeffrey Koplan tried to cut through the bureaucratic red tape surrounding the procurement of smallpox vaccine by calling a meeting in Atlanta of the responsible officials at HHS and Defense and the regulators at the FDA. A key issue was whether the military and civilian contracts should be combined. The Pentagon objected that increasing the number of doses procured from DynPort by an order of magnitude would significantly delay the delivery of the military vaccine. HHS officials also noted that the small-scale production technique that DynPort had chosen for the military contract could not easily be ramped up to satisfy the much larger civilian requirement. Because the company would have to start almost from scratch to modify the manufacturing process, little if any cost savings would result. It was therefore agreed that the CDC should issue a separate contact to the U.S. pharmaceutical industry for the civilian vaccine stockpile.

In mid-2000, the CDC issued a request for proposal to the U.S. pharmaceutical industry seeking competitive bids, and on September 20, the prime contract was awarded to OraVax Corporation of Cambridge, Massachusetts. (OraVax was subsequently acquired by British drugmaker Acambis PLC.) Under this contract, Acambis will develop a cell-culture vaccine based on the same strain of vaccinia as in the old Dryvax vaccine, which was employed successfully in Brazil and many African countries during the WHO eradication campaign. The safety and equivalency of the new vaccine will be tested in five thousand human volunteers by measuring the skin reaction at the vaccination site and the level of antibodies in the recipient's blood, while the protective effect of the vaccine will be tested in monkeys exposed to monkeypox virus. In contrast to Dryvax, the new vaccine will be produced in large-scale bioreactors using advanced cell culture techniques.

Delivery of the new vaccine, initially planned for 2004 or 2005, was moved up to late 2002 after the September 11, 2001, terrorist attacks against New York and Washington made the threat of bioterrorism with smallpox appear more imminent. HHS Secretary Tommy Thompson first increased the Acambis order to 54 million doses of smallpox vaccine. Then, on October 17, he declared that he was negotiating with several drug companies to procure an additional 250 million doses by the end of 2002, although it would not yet be FDA-approved by that date. The result would be a total of 300 million doses of smallpox vaccine—enough for every man, woman, and child in the United States. Once manufactured, the vaccine will be stockpiled at multiple sites around the country, with the locations to be kept secret for security reasons. In the event of a bioterrorist incident involving smallpox, shipments of vaccine will be flown or otherwise transported to the affected area.

Because the new smallpox vaccine will have a shelf life of five years or perhaps longer, depending on stability testing after production, the CDC contract with Acambis calls for annual production of smallpox vaccine through the year 2020 to replace the expired stocks and respond to increased demand as needed. Past experience suggests that if freeze-dried smallpox vaccine is stored in sealed vials at low temperature, it will retain its potency almost indefinitely. Thus, instead of discarding the expired lots of vaccine, the CDC plans to warehouse them indefinitely under optimal storage conditions and to test the expired lots periodically. Over the twenty-year life of the contract, it is anticipated that Acambis will produce and stockpile a total of 168 million doses of smallpox vaccine at a cost of $343 million, or a cost per dose of about $2.

Given that smallpox vaccine can cause serious complications in individuals with eczema or a weakened immune system (as well as in occasional healthy people) and increases the risk of miscarriage during the first trimester of pregnancy, there are no plans to vaccinate the entire U.S. population prophylactically. Instead, the vaccine will be produced and kept on the shelf, for use only in the unlikely event that one or more cases of smallpox are detected and confirmed. One exception is the need to vaccinate scientists and technicians who conduct research with vaccinia and other poxviruses that infect humans. In November 2001, CDC also began vaccinating teams of physicians, epidemiologists, and laboratory workers who would be dispatched to assist local and state health departments if a case of smallpox was suspected.

Beyond the risk of complications, it would be unwise to squander the current supply of vaccine by immunizing the general population prophylactically. U.S. planners worry that the existing stocks of Dryvax could be needed to avert a global pandemic if terrorists were to unleash smallpox outside the United States. Despite the estimated 50 million doses of vaccine stockpiled worldwide, it is far from clear that governments would release their own limited supplies to assist another country. Thus, by "keeping its powder dry," the United States might have a chance to contain a smallpox outbreak overseas. Until the new vaccine becomes available, the U.S. government is exploring ways of stretching the Dryvax stocks. For example, the National Institutes of Health issued a contract to the Center for Vaccine Development at St. Louis University School of Medicine to assess the effect of diluting the Dryvax vaccine fivefold or even tenfold, which could greatly expand the available supply.

Closely related to the stockpile of smallpox vaccine is the dwindling supply of vaccinia immune globulin (VIG), the concentrated solution of antibodies to vaccinia virus that is used to treat the complications of vaccination. The United States currently has only 5,400 vials of VIG, each containing 5 milliliters, or enough altogether for 675 adults. In the past, this amount would have been sufficient to treat the expected rate of complications in 3 million vaccinees. Now, however, the higher incidence of HIV infection and other types of immune system impairment means that the expected rate of complications from smallpox vaccine—and hence the need for VIG—would be significantly greater. Supplies of VIG outside the United States are unknown.

The Pentagon plans to produce more VIG from the frozen blood plasma of volunteers who received multiple smallpox vaccinations in the early 1990s. This product will be more concentrated so that small volumes can be administered intravenously, avoiding the painful process of injecting large amounts of VIG into muscle. Still, there are major hurdles to overcome: Experts worry that the frozen blood plasma could be contaminated with dangerous viruses such as hepatitis C, and because few people are still being vaccinated against smallpox, there is little prospect of obtaining large amounts of VIG-enriched blood plasma in the future. The VIG problem will have to be solved as part of the overall vaccine procurement effort. One possibility might be to develop an anti-vaccinia drug for treating the complications of vaccination in at-risk populations.

The possible return of smallpox is part of the broader threat that infectious diseases pose to global health and security. During the two decades after World War II, new vaccines and antibiotics, combined with better living conditions, dramatically decreased the incidence of the major infectious diseases in industrialized countries. More recently, the rapidly expanding field of molecular biology has provided powerful insights into disease mechanisms and the human immune system, yielding further therapeutic breakthroughs. Nevertheless, it is now widely understood that because microbes are capable of continual evolution, the war against infectious diseases can never be won definitively but must be waged on an ongoing basis.

The unprecedented level of international integration brought about by air travel and free trade has also had the unintended side effect of making the United States far more vulnerable to importations of deadly infections. For example, West Nile encephalitis, a disease indigenous to East Africa and the Middle East, triggered a minor epidemic in New York City in 1999, its first appearance in the Western Hemisphere. Experts speculate that the virus may have been imported into the United States by an infected traveler, a migrating bird, or a stray mosquito on an airplane.

Over the past quarter-century, several previously unknown infections have been identified for which no cures are available, including HIV/AIDS, Ebola, "mad cow disease," hepatitis C, and Nipah virus. Although AIDS was not recognized until the 1980s, it now infects some 36 million people and kills 3 million annually. At the same time, old diseases have staged frightening comebacks. Tuberculosis, malaria, and cholera have reemerged or spread geographically, often in more virulent and antibiotic-resistant forms.

Several factors have facilitated the emergence and spread of infectious diseases. The overprescription and misuse of antibiotics have contributed to a dramatic rise in drug-resistant microbes at a time when the discovery of new antibiotics has lagged; the exploitation of formerly remote jungle areas has brought human populations into contact with numerous exotic viruses; the rise of megacities in the developing world with severe health care deficiencies has created "hot spots" for the evolution of new infectious agents; and the growing volume of cross-border travel and trade caused by economic globalization have provided ample opportunities for microbial "hitchhikers." Viruses also have the potential to mutate into more lethal and contagious forms, as occurred with the Spanish flu, which killed more than 20 million people around the globe in 1918–19 and could well strike again.

Political leaders are gradually recognizing the threat posed by infectious diseases to international health and security. In January 2000, the United Nations Security Council held an unprecedented session devoted exclusively to the social, political, and security consequences of the AIDS pandemic in Africa. Later the same month, the CIA released an unclassified National Intelligence Estimate (NIE) titled *The Global Infectious Disease Threat and Its Implications for the United States*. This landmark study concluded: "New and reemerging infectious diseases will pose a rising global health threat and will complicate U.S. and global security over the next twenty years. These diseases will endanger U.S. citizens at home and abroad, threaten U.S. armed forces deployed overseas, and exacerbate social and political instability in key countries and regions in which the United States has significant interests."

Not only do importations of disease threaten American citizens directly, but devastating epidemics such as AIDS, malaria, and tuberculosis are spawning political turmoil and civil conflict abroad that could ultimately draw in the United States. The NIE predicts that in the hardest-hit countries of the developing and former communist worlds, the persistent burden of infectious disease is likely to aggravate and even provoke social fragmentation and political polarization. AIDS and malaria combined will reduce the GDP of several sub-Saharan African countries by 20 percent or more by 2010, pushing them to the brink of economic collapse as they lose the most productive fraction of the population. If current trends continue, 41.6 million children in twenty-seven countries will have lost one or both parents to AIDS, producing a generation of orphans with little hope of education or employment. These young people may become radicalized or exploited for political ends, as in the increasingly pervasive phenomenon of the child-soldier, putting the affected countries at risk of further economic decline, increased crime, and political instability and violence.

Given this sobering picture, expanded international cooperation in combating emerging and reemerging infections has become essential. In a rapidly shrinking and globalizing world, no country is an island and all are condemned to interdependence. As travel and trade expand, human and animal pathogens will increasingly move across national borders undetected, making any retreat into a "fortress America" illusory. Moreover, unless the United States helps to contain the spread of infectious diseases overseas, the resulting socioeconomic catastrophes could require massive infusions of emergency aid and even the deployment of U.S. military forces to restore order.

Responding to critics who questioned the relevance to national security of infectious disease, President Clinton's national security advisor, Sandy Berger, wrote in *Foreign Affairs*, "[A] problem that kills huge numbers, crosses borders, and threatens to destabilize whole regions is the very definition of a national security threat. . . . To dismiss it as a 'soft' issue is to be blind to hard realities." Prudent investments in international health are needed to avoid future disease-related crises that could prove costly to the United States in both money and lives. Establishing a global infectious disease surveillance and containment system under WHO auspices, for example, would directly benefit the United States as well as other nations.

At the same time that natural emerging infections pose a growing threat to humanity, biological warfare—which has been called "public health in reverse"—is a man-made scourge. Roughly a dozen countries have already acquired or are actively seeking these heinous weapons. Although advances in biotechnology and genetic engineering promise major benefits in medicine and agriculture, they also have a dark side: the potential for developing ever more lethal and insidious instruments of biological warfare.

As these powerful technologies hurtle forward at an accelerating rate, they threaten to outstrip the fragile bulwark of moral, legal, and political constraints designed to prevent their misuse for nefarious purposes. It is to be hoped that in the coming years, the nations of the world will come to view infectious disease as a common enemy and join together to combat the many natural scourges, old and new, that threaten the health and welfare of the human species. Also needed is the international political will to strengthen the existing legal ban on the development and possession of biological weapons by instituting measures for the investigation of suspicious outbreaks of disease and the inspection of dual-use production facilities, such as vaccine plants, backed up with mandatory sanctions on violators. Until humanity's legal and moral restraints catch up with its scientific and technological achievements, the eradication of smallpox will remain as much a cautionary tale as an inspirational one.

Source Notes

One: Monster On Death Row

hundreds of millions of lives: Michael B. A. Oldstone, *Viruses, Plagues, and History* (New York: Oxford University Press, 1998), p. 27.

between ten million and fifteen million people: Donald A. Henderson, "Smallpox—Epitaph for a Killer," *National Geographic*, vol. 154, no. 6 (December 1978), p. 803.

Two: Smallpox And Civilization

jumped the species barrier: Grant McFadden, "What Can Be Learned about the Human Immune System from Variola Virus?" transcript, *Institute of Medicine Workshop on Assessment of Future Needs for Live Variola (Smallpox) Virus*, November 20, 1998 (Washington, D.C.: National Academy of Sciences), p. 25.

small, isolated groups: Jared Diamond, "A Pox upon Our Genes," *Natural History*, vol. 99, no. 2 (February 1990), p. 26.

Ramses V: Donald R. Hopkins, *Princes and Peasants: Smallpox in History* (Chicago: University of Chicago Press, 1983).

Alexander the Great, Marcus Aurelius: Charles Siebert, "Smallpox Is Dead. Long Live Smallpox," *New York Times Magazine*, August 30, 1994, p. 33.

epidemiological adjustment: William H. McNeill, *Plagues and Peoples* (New York: Anchor Books, 1976), p. 144.

diminished psychological burden: McNeill, *Plagues and Peoples*, p. 144.

epidemics in Iceland: Deborah MacKenzie, "Icelanders Argue over Their Ancestors," *New Scientist*, vol. 150, no. 2032 (June 1, 1996), p. 10.

genetic resistance: Diamond, "A Pox upon Our Genes," pp. 28, 30.

powerful advantage: McNeill, *Plagues and Peoples*, p. 86.

slave trade to Hispaniola: Michael B. A. Oldstone, *Viruses, Plagues, and History* (New York: Oxford University Press, 1998), p. 30.

Cortes story, Bernal Diaz quote: Geoffrey Cowley, "The Great Disease Migration," *Newsweek*, special issue, Fall/Winter 1991, p. 54.

psychological impact on Aztecs: McNeill, *Plagues and Peoples*, p. 217.

Pizarro's conquest of Incas: Cowley, "The Great Disease Migration."

Winthrop quote: Corinna Wu, "A Pox on Smallpox," *U.S. News & World Report*, vol. 120, no. 5 (February 5, 1996), p. 14.

Bradford quote: Samuel M. Wilson, "On the Matter of Smallpox," *Natural History*, vol. 103, no. 9 (September 1994), p. 66.

decline of Amerindian population: Russell Thornton, *American Indian Holocaust and Survival* (Norman: University of Oklahoma Press, 1987), pp. 36, 90.

royal deaths: Hopkins, *Princes and Peasants.*

higher education in American colonies: D. A. Henderson, personal communication.

Europe's most devastating disease: Edward S. Golub, *The Limits of Medicine* (New York: Times Books, 1994), p. 114.

epidemics in European cities: Horace G. Ogden, *CDC and the Smallpox Crusade* (Washington, D.C.: U.S. Department of Health and Human Services, HHS Publication No. (CDC) 87–8400, 1988), pp. 3–4.

spread to South Africa and Australia: World Health Organization, *The Global Eradication of Smallpox: Final Report of the Global Commission for the Certification of Smallpox Eradication* (Geneva: WHO, 1980), p. 16.

400,000 Europeans a year: Oldstone, *Viruses, Plagues, and History,* p. 33.

Macaulay quote: Thomas Babington Macaulay, *The History of England from the Accession of James II* (Philadelphia: Claxton, Remsen & Haffelfinger, 1800).

Thackeray quote: William Makepeace Thackeray, *The History of Henry Esmond* (New York: Literary Guild of America, 1950).

smallpox deities: F. Fenner, D. A. Henderson, I. Arita, Z. Jezek, and I. D. Ladnyi, *Smallpox and Its Eradication* (Geneva: World Health Organization, 1988), p. 219.

Rhazes: Nicolau Barquet and Pere Domingo, "Smallpox: The Triumph Over the Most Terrible of the Ministers of Death," *Annals of Internal Medicine,* vol. 127 (October 15, 1997), pp. 635–42.

Queen Elizabeth I: Abbas M. Behbehani, "The Smallpox Story: Life and Death of an Old Disease," *Microbiological Reviews,* vol. 47, no. 4 (December 1983), p. 458.

variolation introduced into China: Yu Xinhua and Yang Qingzhen, eds., *Shengwu Wuqi yu Zhanzheng* [Biological Weapons and Warfare] (Beijing: Guofang Gongye Chubanshe, 1997), pp. 138–39.

insufflation technique: Robert Temple, *The Genius of China: 3000 Years of Science, Discoveries and Inventions* (New York: Simon and Schuster, 1986), pp. 135–37.

Lady Montagu: C. W. Dixon, *Smallpox* (London: J. & A. Churchill, Ltd., 1962), p. 220; Barquet and Domingo, "Smallpox."

Lady Montagu quote: Lady Mary Wortley Montagu, *Letters of the Right Honourable Lady M—y W—y M—e: Written During Her Travels in Europe, Asia, and Africa...,* vol. 1 (Aix: Anthony Henricy, 1796), pp. 167–69; letter 36, to Mrs. S.C. from Adrianople, not dated.

Royal Experiment: Thomas A. Kerns, *Jenner on Trial* (New York: University Press of America, 1997), pp. 8–9.

death of Louis XV: McNeill, *Plagues and Peoples,* p. 256.

Sutton-Dimsdale variolation technique: Donald A. Henderson, "Edward Jenner's Vaccine," *Public Health Reports,* vol. 112, no. 2 (March 13, 1997), p. 116.

variolation in the American colonies: McNeill, *Plagues and Peoples,* p. 257.

Cotton Mather: Wilson, "On the Matter of Smallpox," pp. 66–67.

Benjamin Franklin: Barquet and Domingo, "Smallpox."

military use in French and Indian War: Mark Wheelis, "Biological Warfare Before 1914," in Erhard Geissler and John Ellis van Courtland Moon, *Biological and Toxin Weapons: Research, Development, and Use from the Middle Ages to 1945* (Oxford, England: Oxford University Press for the Stockholm International Peace Research Institute, 1999), pp. 21–24; James A. Poupard, Linda A. Miller, and Lindsay Granshaw, "The Use of Smallpox As a Biological Weapon in the French and Indian War of 1763," *ASM News* [American Society for Microbiology], vol. 55, no. 3 (1989), pp. 122–24.

military use in Revolutionary War: Elizabeth A. Fenn, "Biological Warfare in Eighteenth-Century North America: Beyond Jeffrey Amherst," *Journal of American History,* vol. 86, no. 4 (March 2000), pp. 1552–80; Elizabeth A. Fenn, "Biological Warfare, Circa 1750," *New York Times,* April 11, 1998, p. 11.

THREE: DR. JENNER'S VACCINE

Jenner's observations on cowpox: Thomas A. Kerns, *Jenner on Trial* (New York: University Press of America, 1997), pp. 11–12.

Jenner's letter to friend: Samuel M. Wilson, "On the Matter of Smallpox," *Natural History,* vol. 103, no. 9 (September 1994), p. 64.

satirical cartoon: F. Fenner, D. A. Henderson, I. Arita, Z. Jezek, and I. D. Ladnyi, *Smallpox and Its Eradication* (Geneva: World Health Organization, 1988), p. 269.

second edition of pamphlet: Ola Elizabeth Winslow, *A Destroying Angel: The Conquest of Smallpox in Colonial Boston* (Boston: Houghton Mifflin, 1974), p. 103.

Benjamin Waterhouse: Alvin Powell, "The Beginning of the End of Smallpox," *Harvard University Gazette,* May 20, 1999, pp. 3–4.

transport of cowpox vaccine: Donald A. Henderson, "Edward Jenner's Vaccine," *Public Health Reports,* vol. 112, no. 2 (March 13, 1997), p. 116.

arm-to-arm vaccination: Donald A. Henderson and Frank Fenner, "Smallpox and Vaccinia," in Stanley A. Plotkin and Edward A. Mortimer, Jr., *Vaccines,* 2nd ed. (Philadelphia: W. B. Saunders Co., 1994), pp. 14–15.

de Balmis mission: Wilson, "On the Matter of Smallpox," p. 64; L. Thapalyal, "A Grim Saga," *World Health,* October 1978, p. 6.

Lewis and Clark: David J. Meltzer, "How Columbus Sickened the New World," *New Scientist,* vol. 136, no. 1842 (October 10, 1992), p. 38.

Napoleon Bonaparte: Nicolau Barquet and Pere Domingo, "Smallpox: The Triumph Over the Most Terrible of the Ministers of Death," *Annals of Internal Medicine,* vol. 127 (October 15, 1997), pp. 635–42.

Bleak House: Charles Dickens, *Bleak House* (London: Penguin Classics, 1996), p. 500.

James Smith: Michael B. A. Oldstone, *Viruses, Plagues, and History* (New York: Oxford University Press, 1998), p. 42.

Abraham Lincoln: Fenner et al., *Smallpox and Its Eradication,* p. 240.

military use in Civil War: Jeffrey K. Smart, "History of Chemical and Biological Warfare: An American Perspective," in Frederick R. Sidell, Ernest T. Takafuji, and David R. Franz, eds., *Textbook of Military Medicine, Part I: Warfare, Weaponry, and the Casualty: Medical Aspects of Chemical and Biological Warfare* (Washington, D.C.: Borden Institute, Walter Reed Army Medical Center, 1997), p. 12.

"The Schooner Maryann": W. Roy MacKenzie, *Ballads and Sea Songs from Nova Scotia* (Hatborough, Penn.: Folklore Associates, 1928, reprinted 1963).

Italian children with syphilis: Henderson, "Edward Jenner's Vaccine."

Franco-Prussian War: Henderson, "Edward Jenner's Vaccine."

opposition to vaccination: Michael R. Albert, Kristen G. Ostheimer, and Joel G. Breman, "The Last Smallpox Epidemic in Boston and the Vaccination Controversy, 1901–1903," *New England Journal of Medicine,* vol. 344, no. 5 (February 1, 2001), pp. 375–79.

George Bernard Shaw quote: Derrick Baxby, "The End of Smallpox," *History Today,* March 1999.

nineteenth-century remedies: William G. Eidson, "Confusion, Controversy, and Quarantine: The Muncie Smallpox Epidemic of 1893," *Indiana Magazine of History,* vol. LXXXVI (December 1990), p. 392.

alastrim: Fenner et al., *Smallpox and Its Eradication,* pp. 242–43; Henderson and Fenner, "Smallpox and Vaccinia," p. 29.

mandatory vaccination laws: Horace G. Ogden, *CDC and the Smallpox Crusade* (Washington, D.C.: U.S. Department of Health and Human Services, HHS Publication No. (CDC)87–8400, 1988), pp. 75–76.

incidence in United States: Henderson and Fenner, "Smallpox and Vaccinia," p. 29.

impact of World War II: Abbas M. Behbehani, "The Smallpox Story: Life and Death of an Old Disease," *Microbiological Reviews,* vol. 47, no. 4 (December 1983), p. 482.

1947 outbreak in New York City: Jack W. Hopkins, *The Eradication of Smallpox: Organizational Learning and Innovation in International Health* (Boulder, Colo.: Westview Press, 1989), p. 76.

1949 outbreak in Texas: Ogden, *CDC and the Smallpox Crusade,* p. 76.

Collier vaccine: Henderson, "Edward Jenner's Vaccine."

origins of vaccinia: Peter E. Razzell, *Edward Jenner's Cowpox Vaccine,* 2nd ed., (Firle, Sussex: Caliban Books, 1980); Derrick Baxby, *Jenner's Smallpox Vaccine* (London: Heinemann Educational Books, 1981).

"the annihilation of the Small Pox": Edward Jenner, *The Origin of the Vaccine Inoculation* (London: D. N. Shury, 1801), cited in Fenner et al., *Smallpox and Its Eradication,* p. 261.

FOUR: LAUNCHING THE CRUSADE

Epidemic Intelligence Service: Alexander D. Langmuir and Justin M. Andrews, "Biological Warfare Defense. 2. The Epidemic Intelligence Service of the Communicable Disease Center," *American Journal of Public Health,* vol. 42, no. 3 (March 1952), pp. 235–38.

Fred Soper initiative: Abbas M. Behbehani, "The Smallpox Story: Life and Death of an Old Disease," *Microbiological Review,* vol. 47, no. 4 (December 1983), p. 490.

Brock Chisholm proposal: F. Fenner, D. A. Henderson, I. Arita, Z. Jezek, and I. D. Laynyi, *Smallpox and Its Eradication* (Geneva: World Health Organization, 1988), p. 392.

epidemic in Bombay: Albert Herrlich, "Variola: Eindrücke von einer Epidemie in Bombay im Jahre 1958," *Deutsche medizinische Wochenschrift,* vol. 101 (1958), pp. 1426–1428; translated in F. Fenner et al., *Smallpox and its Eradication,* p. 27.

Viktor Zhdanov speech: Fenner et al., *Smallpox and Its Eradication,* p. 367.

1961–62 outbreak in England and Wales: Derrick Baxby, "The End of Smallpox," *History Today,* March 1999.

suspected importations into U.S.: Horace G. Ogden, *CDC and the Smallpox Crusade* (Washington, D.C.: U.S. Department of Health and Human Services, HHS Publication No. (CDC)87–8400, 1988), pp. 14–15.

complications of vaccination: Fenner et al., *Smallpox and Its Eradication,* p. 299.

Dr. Henry Kempe: Ogden, *CDC and the Smallpox Crusade,* p. 76.

statistics on complications: Derrick Baxby, "Indications for Smallpox Vaccination: Policies Still Differ," *Vaccine,* vol. 11, no. 4 (March 1993), pp. 395–96; Donald

A. Henderson et al., "Smallpox as a Biological Weapon," *Journal of the American Medical Association*, vol. 281, no. 22 (June 9, 1999), p. 2135.

"ped-o-jet" system: Ogden, *CDC and the Smallpox Crusade*, p. 17.

origins of AID/CDC program: Ogden, *CDC and the Smallpox Crusade*, pp. 21–25; Fenner et al., *Smallpox and Its Eradication*, pp. 855, 407–08.

complaint by Soviet delegate: Fenner et al., *Smallpox and Its Eradication*, p. 408.

nearly fifty countries still suffered from smallpox: World Health Organization, *The Global Eradication of Smallpox: Final Report of the Global Commission for the Certification of Smallpox Eradication* (Geneva: World Health Organization, 1980), p. 9.

"Poor countries . . . became disheartened": June Goodfield, *Quest for the Killers* (Boston: Birhaeuser, 1985), p. 194.

Dubos quote: René Dubos, *Man Adapting* (New Haven: Yale University Press, 1965), p. 379.

smallpox elimination from China: Frank Fenner, "The WHO Global Smallpox Eradication Programme: Vaccine Supply and Variola Virus Stocks," in Erhard Geissler and John P. Woodall, eds., *Control of Dual-Threat Agents: The Vaccines for Peace Programme* (Oxford, England: Oxford University Press for the Stockholm International Peace Research Institute, 1994), p. 187.

Henderson meeting with Watt: Goodfield, *Quest for the Killers*, p. 195.

Five: Global Eradication

Henderson meeting with Venediktov: Donald A. Henderson, "Smallpox Eradication: A Saga of Triumph and Betrayal," *Infectious Diseases in Clinical Practice*, vol. 8 (December 1999), p. 446.

Soviet speech at 1971 World Health Assembly: Jack W. Hopkins, *The Eradication of Smallpox: Organizational Learning and Innovation in International Health* (Boulder, Colo.: Westview Press, 1989), p. 32.

quality of vaccine: Frank Fenner, "The WHO Global Smallpox Eradication Programme: Vaccine Supply and Variola Virus Stocks," in Erhard Geissler and John P. Woodall, eds., *Control of Dual-Threat Agents: The Vaccines for Peace Programme* (Oxford, England: Oxford University Press for the Stockholm International Peace Research Institute, 1994), p. 191.

use of Lister strain: World Health Organization, *The Global Eradication of Smallpox: Final Report of the Global Commission for the Certification of Smallpox Eradication* (Geneva: World Health Organization, 1980), p. 27.

invention of bifurcated needle: Joel N. Shurkin, *The Invisible Fire: The Story of Mankind's Victory Over the Ancient Scourge of Smallpox* (New York: G. P. Putnam's Sons, 1979), p. 286; Ogden, *CDC and the Smallpox Crusade*, p. 88.

surveillance-containment strategy: F. Fenner, D. A. Henderson, I. Arita, Z. Jezek, and I. D. Ladnyi, *Smallpox and Its Eradication* (Geneva: World Health Organization, 1988), pp. 493–95.

test of strategy in Eastern Nigeria: author interview with William Foege; Fenner et al., *Smallpox and Its Eradication,* p. 883.

experience in Tamil Nadu, Paraná states: Fenner et al., *Smallpox and Its Eradication,* p. 495.

incorrect assumptions: World Health Organization, *The Global Eradication of Smallpox: Final Report of the Global Commission for the Certification of Smallpox Eradication* (Geneva: World Health Organization, 1980), p. 22.

end of mandatory vaccination in U.S.: Horace G. Ogden, *CDC and the Smallpox Crusade* (Washington, D.C.: U.S. Department of Health and Human Services, HHS Publication No. (CDC)87–8400, 1988), pp. 78–79.

epidemic in Iran, Iraq, Syria: Fenner et al., *Smallpox and Its Eradication,* p. 1089.

outbreak in Yugoslavia: David Egli, "Conquest of an Epidemic," *World Health,* October 1972, pp. 28–30; Fenner et al., *Smallpox and Its Eradication,* pp. 1091–95; D. A. Henderson, "Bioterrorism as a Public Health Threat," *Emerging Infectious Diseases,* vol. 4, no. 3 (July–September 1998), p. 489; Richard Preston, "The Demon in the Freezer," *The New Yorker,* vol. LXXV, no. 18 (July 12, 1999), pp. 48–51.

Six: Long Road To Zero

density and mobility of Indian population: Lawrence B. Brilliant, *The Management of Smallpox Eradication in India* (Ann Arbor: University of Michigan Press, 1985), p. 78; F. Fenner, D. A. Henderson, I. Arita, Z. Jezek, and I. D. Ladnyi, *Smallpox and Its Eradication* (Geneva: World Health Organization, 1988), p. 715.

viral virulence and population density: Wendy Orent, "Killer Pox in the Congo," *Discover,* vol. 20, no. 20 (October 1999), p. 78.

cult of Shitala Mata: Carol Shepherd McLain, "A New Look at an Old Disease: Smallpox and Biotechnology," *Perspectives in Biology and Medicine,* vol. 38, no. 4 (Summer 1995), p. 629; Horace G. Ogden, *CDC and the Smallpox Crusade* (Washington, D.C.: U.S. Department of Health and Human Services, HHS Publication No. (CDC)87–8400, 1988), p. 100.

attitude of middle-class Indians: Jack W. Hopkins, *The Eradication of Smallpox: Organizational Learning and Innovation in International Health* (Boulder, Colo.: Westview Press, 1989), p. 2.

Salt Lake Refugee Camp: A. K. Joarder, D. Tarantola, and J. Tulloch, *The Eradication of Smallpox from Bangladesh* (New Delhi: WHO South-East Asia Regional Office, 1980), pp. 25–30; Fenner et al., *Smallpox and Its Eradication,* p. 747.

Lawrence Brilliant: Lawrence B. Brilliant with Girija Brilliant, "Death for a Killer Disease," *Quest,* May/June 1978, p. 5. In 1978, after the successful conclusion of the smallpox eradication campaign, Brilliant held a meeting at his home in Ann Arbor, Michigan, of several alumni of the WHO country team in India, including Nicole Grasset, Ned Willard, and Steve Jones. Determined to continue their work together in the field of international health, they established the Seva Foundation (www.serva.org) with the aim of reducing the incidence of blindness in India and Nepal. By funding hospitals for cataract surgery, Seva has since helped more than 2 million blind people to recover their sight.

"active search" strategy: Fenner et al., *Smallpox and Its Eradication,* pp. 753–762; R. N. Basu, Z. Jezek, and N. A. Ward, *The Eradication of Smallpox from India* (New Delhi: WHO South-East Asia Regional Office, 1979), pp. 29–31.

Tatanagar epidemic: Brilliant, *The Management of Smallpox Eradication in India,* pp. 46–50; Fenner et al., *Smallpox and Its Eradication,* pp. 770–72; Brilliant, "Death for a Killer Disease," pp. 5, 9.

"house on fire" analogy: Fenner et al., *Smallpox and Its Eradication,* p. 767.

Mohan Singh episode: Brilliant, "Death for a Killer Disease," p. 5.

Operation Smallpox Zero: Basu, Jezek, and Ward, *The Eradication of Smallpox from India,* p. 32.

reward program: Hopkins, *The Eradication of Smallpox,* p. 100; Ogden, *CDC and the Smallpox Crusade,* p. 104.

smallpox in Bangladesh: Joarder, Tarantola, and Tulloch, *The Eradication of Smallpox from Bangladesh,* pp. 27–30; June Goodfield, *Quest for the Killers* (Boston: Birhaeuser, 1985), pp. 225–28; Hopkins, *The Eradication of Smallpox,* p. 92; Fenner et al., *Smallpox and Its Eradication,* pp. 807–43.

outbreak on Bhola Island: Goodfield, *Quest for the Killers,* pp. 234–37; Joarder, Tarantola, and Tulloch, *The Eradication of Smallpox from Bangladesh,* pp. 38–48; Fenner et al., *Smallpox and Its Eradication,* pp. 843–47.

geography of Ethiopia: Marcus Brooke, "Ethiopia: Great Expectations," *World Health,* October 1972, p. 24.

WHO program in Ethiopia: Fenner et al., *Smallpox and Its Eradication,* pp. 1018–35.

nomads in Somalia: Z. Jezek, M. Al Aghbari, R. Hatfield, and A. Deria, *Smallpox Eradication in Somalia* (WHO Regional Office for the Eastern Mediterranean, 1979), pp. 77–84.

Ali Maow Maalin: Fenner et al., *Smallpox and Its Eradication,* pp. 1062–63.

telegram to Geneva: Donald A. Henderson, "Smallpox—Epitaph for a Killer?" *National Geographic,* vol. 154, no. 6 (December 1978), p. 797.

SEVEN: REALM OF THE FINAL INCH

certification process: Jack W. Hopkins, *The Eradication of Smallpox: Organizational Learning and Innovation in International Health* (Boulder, Colo.: Westview Press, 1989), p. 124; Lawrence B. Brilliant, *The Management of Smallpox Eradication in India* (Ann Arbor: University of Michigan Press, 1985), p. 68.

"realm of the final inch": Alexander Solzhenitsyn, *The First Circle* (New York: Bantam Books, 1976), p. 161.

Global Commission: World Health Organization, *The Global Eradication of Smallpox: Final Report of the Global Commission for the Certification of Smallpox Eradication* (Geneva: World Health Organization, 1980), p. 9.

WHO laboratory survey: D. A. Henderson, "Case Studies of Major Eradication Efforts: Smallpox," presentation at Institute of Medicine, Forum on Emerging Infections, *The Consequences of Viral Disease Eradication: Addressing Post-Immunization Challenges,* Washington, D.C., February 1, 2001.

accident at London School: F. Fenner, D. A. Henderson, I. Arita, Z. Jezek, and I. D. Ladnyi, *Smallpox and Its Eradication* (Geneva: World Health Organization, 1988), pp. 1095–96.

Birmingham outbreak: Lawrence McGinty, "Smallpox Laboratories, What Are the Risks?" *New Scientist,* vol. 81, no. 1136 (January 4, 1979), pp. 8–14; Fenner et al., *Smallpox and Its Eradication,* pp. 1097–98; Lawrence K. Altman, "Criticism is Leveled in Aftermath of Fatal British Smallpox Outbreak," *New York Times,* February 11, 1979, p. 34.

"Abid" strain: Abbas M. Behbehani, "The Smallpox Story: Life and Death of an Old Disease," *Microbiological Reviews,* vol. 47, no. 4 (December 1983), p. 502.

Bedson remark on August 31: Associated Press report, September 6, 1978.

Bedson suicide: William Stockton, "Smallpox Is Not Dead," *New York Times Magazine,* February 4, 1979, p. 36.

Shooter committee findings: R. A. Shooter et al., *Report of the Investigation Into the Cause of the 1978 Birmingham Smallpox Occurrence* (London: Her Majesty's Stationery Office, July 1980), pp. 2, 38, 60–62.

Meschede incident: D. A. Henderson, "Bioterrorism as a Public Health Threat," *Emerging Infectious Diseases,* vol. 4, no. 3 (July–September 1998), p. 490; Fenner et al., *Smallpox and Its Eradication,* p. 193.

debunking of whitepox: Fenner et al., *Smallpox and Its Eradication,* p. 1333.

cost of WHO campaign: June Goodfield, *Quest for the Killers* (Boston: Birhaeuser, 1985), p. 242.

number of international advisors: Fenner et al., *Smallpox and Its Eradication,* pp. 443–44.

transfer of British stocks to the CDC: Jolyon Jenkins, reporter, and Sarah Lewthwaite, producer, "Killing the Killers," BBC Radio 4, broadcast on May 12, 1999; Robin Marantz Henig, *A Dancing Matrix: Voyages Along the Viral Frontier* (New York: Alfred A. Knopf, 1993), p. 239.

destruction of South African stocks: Fenner et al., *Smallpox and Its Eradication,* p. 1275.

CDC samples from army, ATCC: Nicholas Wade, "Biological Warfare Fears May Impede Last Goal of Smallpox Eradicators," *Science,* vol. 201, no. 4353 (July 28, 1978), pp. 329–30.

contents of Russian repository: L. S. Sandakhchiev, S. S. Marennikova, and A. A. Guskov, "Status of Variola Virus Stock and Complete Inventory of Isolates at WHO Collaborating Centre, Koltsovo," paper delivered at meeting of WHO Ad Hoc Committee on Orthopoxvirus Infections, Geneva, Switzerland, January 14–15, 1999, p. 3.

discovery of smallpox stocks in California: Carol Shepherd McLain, "A New Look at an Old Disease: Smallpox and Biotechnology," *Perspectives in Biology and Medicine,* vol. 38, no. 4 (Summer 1995), p. 632.

discovery of smallpox stocks in Prague: "An Order Was Issued to Destroy the Smallpox Virus; However, It Was Probably Never Carried Out," Prague, *Mlada Fronta Dnes,* in Czech, May 2, 1994, p. 2; translated by Foreign Broadcast Information Service, JPRS-TAC-94-005–L, May 25, 1994, pp. 3–4.

vaccination of U.S. military personnel: Major General V. I. Evstigneev, "The Contemporary State of Vaccinal Prevention of Natural Smallpox in the Armed Forces of Capitalist States," *Voyenno-Meditsinskiy Zhurnal,* in Russian, no. 11, November 1991, pp. 68–70; translated by Foreign Broadcast Information Service, JPRS-ULS-92-025, November 25, 1992, pp. 20–22.

Burgasov statement: Linnea Capps, Sten H. Vermund, and Christine Johnsen, "Smallpox and Biological Warfare: The Case for Abandoning Vaccination of Military Personnel," *American Journal of Public Health,* vol. 76, no. 10 (October 1986), p. 1230.

New York Times editorial: "End the Fear of Smallpox," *New York Times,* January 30, 1984, p. A16.

vaccination of U.S. special forces: McLain, "A New Look at an Old Disease," p. 633.

EIGHT: THE SOVIET BETRAYAL

U.S., Canadian, British studies in World War II: Ed Regis, *The Biology of Doom: The History of America's Secret Germ Warfare Project* (New York: Henry Holt & Co., 1999), p. 219.

Japanese experiments in Manchuria: Robert Harris and Jeremy Paxman, *A Higher Form of Killing: The Secret Story of Gas and Germ Warfare* (London: Chatto & Windus, 1982),

p. 79; Yu Xinhua and Yang Qingzhen, eds., *Shengwu Wuqi yu Zhanzheng* [Biological Weapons and Warfare], (Beijing: Guofang Gongye Chubanshe, 1997), p. 37.

Center of Virology in Zagorsk: Fedor Smirnov, "Taming Viruses: Center for Special Diagnosis and Treatment of Ultradangerous and Exotic Infectious Diseases," *Meditsinskaya Gazeta*, in Russian, December 29, 1999, pp. 10–11.

India-1967 strain: Ken Alibek with Stephen Handelman, *Biohazard: The Chilling True Story of the Largest Covert Biological Weapons Program in the World—Told from Inside by the Man Who Ran It* (New York: Random House, 1999), p. 112.

mass production at Zagorsk: "Biological Weapons in the Former Soviet Union: An Interview with Dr. Kenneth Alibek," *The Nonproliferation Review*, vol. 6, no. 3 (Spring–Summer 1999), p. 3.

mobilization line at Pokrov: Alibek with Handelman, *Biohazard*, p. 112.

single-warhead missiles in Arctic: Author interview with Kenneth Alibek.

Soviet military doctrine: "Biological Weapons in the Former Soviet Union," p. 3.

Yuri Ovchinnikov: Christopher J. Davis, "Nuclear Blindness: An Overview of the Biological Weapons Programs of the Former Soviet Union and Iraq," *Emerging Infectious Diseases*, vol. 5, no. 4 (July–August 1999), p. 510.

U.S. deception operations: David Wise, *Cassidy's Run: The Secret Spy War Over Nerve Gas* (New York: Random House, 2000); Raymond L. Garthoff, "Polyakov's Run," *Bulletin of the Atomic Scientists*, vol. 56, no. 5 (September/October 2000), pp. 37–40.

Zhdanov involvement in Biopreparat: Wendy Orent, "After Anthrax," *The American Prospect*, May 8, 2000, p. 19.

radiation-resistant additives: Author interview with former Vector scientist.

Alibekov's education: Tom Mangold and Jeff Goldberg, *Plague Wars: A True Story of Biological Warfare* (New York: St. Martin's Press, 1999), p. 184.

1986–90 five-year-plan: Alibek with Handelman, *Biohazard*, p. 118.

4,500 scientists at Vector: Sergey V. Netesov, "The Scientific and Production Association Vector: The Current Situation," in Erhard Geissler and John P. Woodall, eds., *Control of Dual-Threat Agents: The Vaccines for Peace Programme* (Oxford, England: Oxford University Press for the Stockholm International Peace Research Institute, 1994), p. 133.

description of Vector in Russian press: Vladimir Petrovskiy, "National 'Virus Property,'" *Obshchaya Gazeta*, in Russian, no. 18 (May 8–14, 1997), p. 10; translated in Foreign Broadcast Information Service, FBIS-UST-97-019, May 14, 1997.

Lev Sandakhchiev: Author interview with former Vector scientist.

Building 6 at Vector: Sergey V. Netesov and Lev S. Sandakhchiev, "The Development of a Network of International Centers to Combat Infectious Diseases and

Bioterrorism Threats," *The ASA Newsletter* [Applied Science and Analysis, Inc.], February 19, 1999, p. 5; author interview with Peter Jahrling; Mangold and Goldberg, *Plague Wars*, p. 132.

"crocodiles": Author interview with David Franz.

Lukin and Stavsky: Author interviews with Alibek and a former Vector scientist.

production of smallpox at Vector: Alibek with Handelman, *Biohazard*, pp. 118–22.

biological warheads for SS-18: Alibek with Handelman, *Biohazard*, p. 5; Mangold and Goldberg, *Plague Wars*, p. 83; author interview with Alibek.

use of vaccinia as surrogate: Ken Alibek, "Behind the Mask: Biological Warfare," *Perspective* [Institute for the Study of Conflict, Ideology and Policy], vol. IX, no. 1 (September-October 1998); "Biological Weapons in the Former Soviet Union," p. 8.

smallpox chimeras: Wendy Orent, "Escape from Moscow," *The Sciences*, May/June 1998, p. 29; Alibek with Handelman, *Biohazard*, pp. 259–60.

Okhotnik **program:** Author interviews with Sergei Popov and a former Vector scientist.

"absolute" weapon: On-camera interview with Dr. Kanatjan Alibekov, Public Broadcasting System, *Frontline: Plague War*, broadcast on October 13, 1998.

U.S.-British inspection at Vector: Richard Preston, "The Demon in the Freezer," *The New Yorker*, vol. LXXV, no. 18 (July 12, 1999) p. 58; Mangold and Goldberg, *Plague Wars*, pp. 134–36.

expeditions to Russian Arctic: Petrovskiy, "National 'Virus Property,'" p. 10; Judith Miller, "U.S. Helps Russia Turn Germ Center to Peace Uses," *New York Times*, January 8, 2000, p. A3; Orent, "Escape from Moscow," p. 26.

NINE: STAY OF EXECUTION

Louis Sullivan speech: Brian W. J. Mahy, Joseph J. Esposito, and J. Craig Venter, "Sequencing the Smallpox Virus Genome," *ASM News* [American Society for Microbiology], vol. 57, no. 11 (1991), p. 577.

memorandum of understanding: "Memorandum of Understanding between the Department of Defense and the Department of Health and Human Services on Variola Reference Strains and Smallpox Vaccine Stockpile," signed by representatives of both departments on October 21 and December 23, 1991.

DNA sequencing: David Brown, "Computers to Hold Vestiges of Smallpox," *Washington Post*, May 11, 1992, p. A3; Mahy, Esposito, and Venter, "Sequencing the Smallpox Virus Genome," pp. 577–80.

Russian sequencing of India-1967: S. N. Shchelkunov, et al., "Study of Structural-Functional Organization of the Genome of Smallpox Virus. V. Sequencing and Analysis of the Sequence of Nucleotides of the Left End of the Genome of Strain

India-1967," *Molekulyarnaya Biologiya*, in Russian, vol. 30, no. 3 (May–June 1996), pp. 595–612.

financial hardship at Vector: Sergey V. Netesov, "The Scientific and Production Association Vector: The Current Situation," in Erhard Geissler and John P. Woodall, eds., *Control of Dual-Threat Agents: The Vaccines for Peace Programme* (Oxford, England: Oxford University Press for the Stockholm International Peace Research Institute, 1994), p. 137.

Glasgow conference: Lawrence K. Altman, "Scientists Debate Destroying the Last Strains of Smallpox," *New York Times*, August 30, 1993, p. 1; transcript of roundtable discussion, "Smallpox: The Final Steps Toward Eradication," Ninth International Congress of Virology, Glasgow, Scotland, August 11, 1993.

destructionist arguments: Brian W. J. Mahy et al., "The Remaining Stocks of Smallpox Virus Should Be Destroyed," *Science*, vol. 262 (November 19, 1993), pp. 1223–24.

Andzhaparidze quote: Charles Siebert, "Smallpox Is Dead. Long Live Smallpox," *New York Times Magazine*, August 30, 1994.

retentionist arguments: Wolfgang K. Joklik et al., "Why the Smallpox Virus Stocks Should Not Be Destroyed," *Science*, vol. 262 (November 19, 1993), pp. 1225–26; Wolfgang K. Joklik, "The Remaining Smallpox Virus Stocks Are Too Valuable to Be Destroyed," *The Scientist*, vol. 10, no. 24 (December 9, 1996), p. 11; Myrna E. Watanabe, "Poxvirus Research Advances May Stay Stock Destruction," *The Scientist*, vol. 8, no. 4 (February 21, 1994), p. 16.

smallpox as scientific treasure trove: R. F. Massung et al., "Potential Virulence Determinants in Terminal Regions of Variola (Smallpox) Virus Genome," *Nature*, vol. 366 (1993), pp. 748–51; Mark Buller, "How Can Variola Contribute to Understanding Viral Pathogenesis?" presentation at Institute of Medicine workshop, *Assessment of Future Needs for Live Variola (Smallpox) Virus*, November 20, 1998, National Academy of Sciences, Washington, D.C., transcript, p. 6.

deliberate extinction: F. Fenner, D. A. Henderson, I. Arita, Z. Jezek, and I. D. Ladnyi, *Smallpox and Its Eradication* (Geneva: World Health Organization, 1988), p. 1339.

Ghendon quote: Altman, "Scientists Debate Destroying Last Strains of Smallpox."

Dowdle quote: Lawrence K. Altman, "Smallpox Virus, Frozen in 2 Labs, Escapes a Scalding End for Now," *New York Times*, December 25, 1993, p. 1.

Lee letter: Letter from Philip R. Lee, Assistant Secretary for Health, U.S. Department of Health and Human Services, to Dr. Hiroshi Nakajima, Director-General, World Health Organization, December 23, 1993.

Moscow repository during 1993 unrest: Siebert, "Smallpox Is Dead. Long Live Smallpox."

transfer of virus stocks to Vector: L. S. Sandakhchiev, S. S. Marennikova, and A. A. Guskov, "Status of Variola Virus Stock and Complete Inventory of Isolates at WHO Collaborating Centre, Koltsovo," paper delivered at meeting of WHO Ad Hoc Committee on Orthopoxvirus Infections, Geneva, Switzerland, January 14–15, 1999.

smallpox repository at Vector: Vladimir Petrovskiy, "National 'Virus Property,'" *Obshchaya Gazeta,* no. 18 (May 8–14, 1997), p. 10; translated by Foreign Broadcast Information Service, FBIS-UST-97–019, May 14, 1997.

security measures at Vector: E-mail message from Sergei V. Netesov to D. A. Henderson, April 2, 1998.

meeting of WHO Committee: World Health Organization, "Report of the Meeting of the *Ad Hoc* Committee on Orthopoxvirus Infections, Geneva, Switzerland, 9 September 1994," WHO document no. WHO/CDS/BVI/94.3.

Fenner quote: Lawrence K. Altman, "Destruction of Smallpox Virus Backed in WHO Committee," *New York Times,* September 10, 1994, p. 5.

U.S. policymaking process: Author interviews with Henderson, Harris, Kadlec, Lederberg, Miller.

January 1995 meeting of WHO Executive Board: John Maurice, "Virus Wins Stay of Execution," *Science,* vol. 267 (January 27, 1995), p. 450.

WHO Assistant Director-General quote: Lawrence K. Altman, "Lab Samples of Smallpox Win Reprieve," *New York Times,* January 19, 1995, p. 15.

Joseph's response to Henderson: Letter from Dr. Stephen C. Joseph, Assistant Secretary of Defense for Health, to D. A. Henderson, dated February 23, 1995.

Joint Coordinating Group: Author interviews with D. A. Henderson.

change in HHS position: "Smallpox Review," memorandum from Philip R. Lee, M.D., Assistant Secretary for Health, HHS, for Mr. Daniel B. Poneman, Special Assistant to the President and Senior Director for Nonproliferation and Export Controls, NSC staff, April 1995.

Pentagon decision to retain virus: "Smallpox Research Report," memorandum from Stephen C. Joseph, M.D., and Philip R. Lee, M.D., for Mr. Daniel B. Poneman, Special Assistant to the President and Senior Director for Nonproliferation and Export Controls, NSC staff, December 15, 1995.

Henderson was outraged: Handwritten fax from D. A. Henderson to Dr. Robert Knouss, U.S. Public Health Service, December 13, 1995.

President Clinton's decision: "Guidance Paper for 97th WHO Executive Board, Communicable Disease Prevention and Control: Smallpox Eradication—Destruction of Variola Virus Stock (Agenda Item 7.1)," undated document.

January 1996 WHO Executive Board meeting: U.S. Department of State reporting cable, GENEVA 00536, "WHO Recommends Destruction of Smallpox Stocks in 1999," dated January 24, 1996; "Truly an End to Smallpox," *Science News,* vol. 149, no. 23 (June 8, 1996), p. 367.

TEN: WASHINGTON RECONSIDERS

WHO survey: World Health Organization, "Smallpox Eradication—Destruction of Variola Virus Stocks," WHO document no. C.L.3.1998, dated February 2, 1998.

results of WHO survey: David Brown, "Smallpox's Threat as Weapon Is Weighed," *Washington Post,* March 15, 1999, p. A1.

Alibekov goes public: Tim Weiner, "Soviet Defector Warns of Biological Weapons," *New York Times,* February 25, 1998, pp. A1, A8; Richard Preston, "Annals of Warfare: The Bioweaponeers," *New Yorker,* vol. LXXIV, no. 3 (March 9, 1998), pp. 52–65; Ken Alibek, "Russia's Deadly Expertise," *New York Times,* March 27, 1998, p. 19; William J. Broad and Judith Miller, "Soviet Defector Says China Had Accident at a Germ Plant," *New York Times,* April 5, 1999, p. A3.

World Trade Center bombing: John V. Parachini, "The World Trade Center Bombers (1993)," in Jonathan B. Tucker, ed., *Toxic Terror: Assessing Terrorist Use of Chemical and Biological Weapons* (Cambridge, Mass.: The MIT Press, 2000), pp. 185–206.

Aum Shinrikyo: David Kaplan, "Aum Shinrikyo (1995)," in Tucker, ed., *Toxic Terror,* pp. 207–26.

Clinton reaction to novel: William J. Broad and Judith Miller, "Germ Defense Plan in Peril as Its Flaws Are Revealed," *New York Times,* August 7, 1998, p. 1; Stephen S. Hall, "Science-Fiction Policy," *Technology Review,* November/December 1998, p. 92.

Larry Wayne Harris: Jessica Stern, "Larry Wayne Harris (1998)," in Tucker, ed., *Toxic Terror,* pp. 227–46.

Blair House exercise: Judith Miller and William J. Broad, "Exercise Finds U.S. Unable to Handle Germ War Threat," *New York Times,* April 26, 1998, p. A1; Ehud Sprinzak, "The Great Super-Terrorism Scare," *Foreign Policy,* Fall 1998, p. 110.

White House briefing: Broad and Miller, "Germ Defense Plan in Peril as Its Flaws Are Revealed"; Lois R. Ember, "Bioterrorism: Combating the Threat," *Chemical & Engineering News,* July 5, 1999, p. 13.

sixteen-page letter report: Bradley Graham, "Clinton Calls for Germ War Antidotes," *Washington Post,* May 21, 1998, p. 1.

three-quarters of panel's recommendations: Ember, "Bioterrorism," p. 15.

Clinton speech at Annapolis: The White House, Office of the Press Secretary, "Remarks by the President at the United States Naval Academy Commencement," May 22, 1998; Jonathan Peterson, "Clinton Plans Vaccine Stockpile Against Germ Terrorism," *Los Angeles Times,* May 21, 1998, p. 15.

new initiative: The White House, Office of the Press Secretary, "Fact Sheet: Keeping America Secure for the 21st Century: President Clinton's Initiative on Biological and Chemical Weapons Preparedness," January 22, 1999.

observers expressed surprise: Graham, "Clinton Calls for Germ War Antidotes."

"quiet war" had broken out: Broad and Miller, "Germ Defense Plan in Peril."

Clinton remarks on Bin Laden: Judith Miller and William J. Broad, "Clinton Describes Terrorism Threat for 21st Century," *New York Times,* January 22, 1999, p. 1.

Henderson views on smallpox threat: D. A. Henderson, "Smallpox: Clinical and Epidemiological Features," *Emerging Infectious Diseases,* vol. 5, no. 4 (July–August 1999), pp. 537–39; Donald A. Henderson, "The Looming Threat of Bioterrorism," *Science,* vol. 283 (February 26, 1999), p. 1281; Joel G. Breman and D. A. Henderson, "Poxvirus Dilemmas—Monkeypox, Smallpox, and Biologic Terrorism," *New England Journal of Medicine,* vol. 339, no. 8 (August 20, 1998), pp. 556–59.

Ken Bernard: Author interview with Ken Bernard.

Interagency Working Group: Lawrence K. Altman, "Killer Smallpox Gets a New Lease on Life," *New York Times,* May 25, 1999, p. 3.

Huggins presentation at IOM workshop: John Huggins, "Strategies and Progress in Development of Antiviral Therapies," in Institute of Medicine, *Workshop: Assessment of the Future Scientific Needs for Live Variola (Smallpox) Virus,* November 20, 1998, Washington, D.C., transcript, pp. 75, 90–91.

vaccine complications in HIV-infected soldier: Robert R. Redfield, D. Craig Wright, William D. James, T. Stephen Jones, Charles Brown, and Donald S. Burke, "Disseminated Vaccinia in a Military Recruit with Human Immunodeficiency Virus (HIV) Disease," *New England Journal of Medicine,* vol. 316, no. 11 (March 12, 1987), pp. 673–75.

secret report leaked to *New York Times*: William J. Broad and Judith Miller, "Government Report Says 3 Nations Hide Stocks of Smallpox," *New York Times,* June 13, 1999, pp. A1, A12.

concerns about Russia: Esther B. Fein, "The Shots Heard 'Round the World'," *New York Times,* December 21, 1997, p. 12; Broad and Miller, "Government Report Says 3 Nations Hide Stocks of Smallpox"; David Brown, "Smallpox's Threat as Weapon is Weighed," *Washington Post,* March 15, 1999, p. A1.

Soviet scientists in Iran: Judith Miller and William J. Broad, "Iranians, Bioweapons in Mind, Lure Needy Ex-Soviet Scientists," *New York Times,* December 8, 1998, p. A1.

Alibekov quote: Elizabeth Kaledin, correspondent, "Smallpox: A Military Threat," *CBS News,* June 24, 1999, 10:13 P.M. EST.

Russian report on North Korea: Russian Federation, Foreign Intelligence Service, *A New Challenge After the Cold War: Proliferation of Weapons of Mass Destruction,* translation by Foreign Broadcast Information Service, JPRS-TND-93–007, March 5, 1993, p. 30.

blood samples from North Korean soldiers: Broad and Miller, "Government Report Says 3 Nations Hide Stocks of Smallpox."

possible sources of North Korean stocks: Joseph S. Bermudez, Jr., "Exposing the North Korean BW Arsenal," *Jane's Intelligence Review,* August 1998, p. 28.

Iraqi smallpox vaccination: U.S. Defense Intelligence Agency, "Immunization of Iraqi Military Personnel/Desert Shield," declassified intelligence report no. IIR 1 521 0157 91, Department of Defense GulfLink web site; Defense Intelligence Agency, "AFMIC Assessment Based on Laboratory Analysis of Iraqi Serum Samples by U.S. Army Medical Research Institute for [*sic*] Infectious Diseases," Department of Defense GulfLink web site, filename 0pgv0014.oop.

Iraqi research on camelpox: Broad and Miller, "Government Report Says 3 Nations Hide Stocks of Smallpox."

human experimentation with camelpox: Marie Colvin and Uzi Mahnaimi, "Saddam Tested Anthrax on Human Guinea Pigs," *The Sunday Times* [London], January 18, 1998, Foreign News, p. 1.

freeze-drier labeled "smallpox": Broad and Miller, "Government Report Says 3 Nations Hide Stocks of Smallpox."

alleged Soviet transfer of virus to Iraq: U.S. Defense Intelligence Agency, "Russian Biological Warfare Technology," declassified intelligence report IIR 2 201 0910 94, dated May 20, 1994, Department of Defense GulfLink web site.

sighting of Vector scientist in Baghdad: Tom Mangold and Jeff Goldberg, *Plague Wars: A True Story of Biological Warfare* (London: Macmillan, 1999), p. 288.

other countries with undeclared stocks: Richard Preston, "The Demon in the Freezer," *The New Yorker,* vol. LXXV, no. 18 (July 12, 1999), p. 46; Agence France Presse, "Israel Holding Smallpox Virus Stocks in Violation of Accords: Report," January 28, 1999, 08:25 GMT.

"perpetuating a fraud": Broad and Miller, "Government Report Says 3 Nations Hide Stocks of Smallpox."

Clinton speech at National Academy: The White House, Office of the Press Secretary, "Remarks by the President on Keeping America Secure for the 21st Century," National Academy of Sciences, Washington, D.C., January 22, 1999.

Clinton statement on IOM report: Judith Miller, "Panel Says Smallpox Stocks May Be Useful," *New York Times,* May 16, 1999, p. A8.

Eleven: Decision In Geneva

release of IOM report: Paul Recer, "Save the Smallpox? Experts Want Some of Live Virus for Study," Associated Press, March 15, 1999.

main findings of IOM report: Institute of Medicine, *Assessment of Future Scientific Needs for Live Variola Virus* (Washington, D.C.: National Academy Press, 1999), pp. 2–4.

Carpenter quote: Jolyon Jenkins, reporter, and Sarah Lewthwaite, producer, "Killing the Killers," BBC Radio 4, broadcast, May 12, 1999.

"Science is science": Judith Miller, "Panel Says Smallpox Stocks May Be Useful," *New York Times,* March 16, 1999, p. A8.

letter from seven Republican senators: Letter to President Bill Clinton signed by U.S. Senators John Kyl, Jesse Helms, Trent Lott, Richard Lugar, Frank Murkowski, Craig Thomas, and Paul Coverdale, dated March 22, 1999.

official White House statement: The White House, Office of the Press Secretary, "White House Statement on Destruction of Stocks of Smallpox Virus," U.S. Newswire, April 22, 1999, 7:55 P.M.; Susan Okie, "U.S. to Oppose Destroying Smallpox Stocks," *Washington Post,* April 23, 1999, p. A2.

joint research with Russia: Judith Miller, "U.S. Foresees Smallpox Research With Russia," *New York Times,* April 23, 1999, p. 23.

Henderson commentary: D. A. Henderson, "Smallpox Reversal Could Be Dangerous," *Baltimore Sun,* May 6, 1999, p. 25A.

Bernard strategy: Author interview with Ken Bernard.

text of draft resolution: World Health Organization, Fifty-second World Health Assembly, Agenda Item 13, "Smallpox Eradication: Destruction of Variola Virus Stocks," A52.A/Conference paper No. 1, May 20, 1999.

WHO decision: Judith Miller and Lawrence K. Altman, "Health Panel Recommends a Reprieve for Smallpox," *New York Times,* May 22, 1999, p. 3; Elizabeth Olson, "WHO Defers Destroying Stocks of Smallpox," *International Herald Tribune,* May 25, 1999, p. 10; Lawrence K. Altman, "Killer Smallpox Gets a New Lease on Life," *New York Times,* May 25, 1999, p. D3; Susan Okie, "Countries Hold Off on Destroying Smallpox Stocks," *Washington Post,* May 25, 1999, p. A2.

Heymann quote: Jenkins and Lewthwaite, "Killing the Killers."

Alibek op-ed: Ken Alibek and Stephen Handelman, "Smallpox Could Still Be a Danger," *New York Times,* May 24, 1999, p. 27.

WHO Advisory Committee: World Health Organization, "Smallpox Eradication: WHO Advisory Committee on Variola Virus Research," *Weekly Epidemiological Record,* vol. 75, no. 6 (February 11, 2000), pp. 45–48; WHO Advisory Committee on Variola Virus Research, "Report of a WHO Meeting, Geneva, Switzerland, 6–9 December 2000," Report no. WHO/CDS/2000.1.

foot-dragging at CDC: Author interview with Jim Hughes.

smallpox research agenda: Donna E. Shalala, "Smallpox: Setting the Research Agenda," *Science,* vol. 285 (August 13, 1999), p. 1011.

smallpox research at Vector: U.S. General Accounting Office, "Biological Weapons: Effort to Reduce Former Soviet Threat Offers Benefits, Poses New Risks," GAO-NSAID-00–138, April 2000, p. 28.

TWELVE: THE UNFINISHED CONQUEST

U.S. Army field test in 1965: William J. Broad, "Smaller, Cheaper, Stealthier, Deadlier," *New York Times* [Week in Review], February 11, 2001, p. 18.

terrorist attack scenario: Tara O'Toole, "Smallpox: An Attack Scenario," *Emerging Infectious Diseases,* vol. 5, no. 4 (July–August 1999), pp. 540–46; Laurie Garrett, "Countries Lose Out in Germ Terror Test," *Newsday,* February 22, 1999, p. 4A.

stockpile of Dryvax vaccine: Associated Press, "Pennsylvania Town Has the Nation's Only Stockpile of Smallpox Vaccine," July 21, 1999; Steve Goldstein, "Old Scourge Kindles Fear of Biological Terrorism," *Philadelphia Inquirer,* April 2, 2000, p. 1.

vaccine still potent: James W. LeDuc and John Becher, "Current Status of Smallpox Vaccine," *Emerging Infectious Diseases,* vol. 5, no. 4 (July–August 1999), p. 593.

smallpox vaccine stocks: Lawrence K. Altman, "Smallpox Vaccine Urged to Prepare for Terrorist Attacks," *New York Times,* March 11, 1998, p. 21.

50 million doses worldwide: D. A. Henderson, "Risk of a Deliberate Release of Smallpox Virus: Its Impact on Virus Destruction," paper presented at meeting of the WHO Ad Hoc Committee on Orthopoxvirus Infections, Geneva, January 14–15, 1999, p. 4.

Pentagon contract with DynPort: U.S. Army, "Army Pursues Joint Vaccine Acquisition Program," Press Release No. 97–59 (Washington, D.C.: U.S. Army, Office of Public Affairs, June 12, 1997).

DynPort subcontract: Dana Hedgpeth, "BioReliance vs. Bioterrorism," *Washington Post*, August 24, 2000, p. E01.

time-frame for acquiring vaccine: "Is Smallpox History?" *The Lancet*, vol. 353, no. 9164 (May 8, 1999), p. 1539.

$77 a dose: Author interview with D. A. Henderson.

vaccine meeting at CDC: Richard Preston, "Updating the Smallpox Vaccine," *The New Yorker*, vol. LXXV, no. 42 (January 17, 2000), p. 27; Donald A. Henderson, Thomas V. Inglesby, John G. Bartlett, et al., "Smallpox as a Biological Weapon," *Journal of the American Medical Association*, vol. 281, no. 22 (June 9, 1999), p. 2136.

OraVax contract: "OraVax in $343m Contract: To Develop Smallpox Vaccine for Government," *Boston Globe*, September 21, 2000, p. C1; Peptide Therapeutics Group Plc, "Award of Major Contract for Smallpox Vaccine from US CDC (Centers for Disease Control and Prevention)," press release, September 20, 2000; Elizabeth Neus, "U.S. Will Stockpile a Defense Against Smallpox," *USA Today/Gannett News Service*, October 11, 2000, p. 6D; Raja Mishra, "Cambridge Firm Accelerates Vaccine Production," *Boston Globe*, October 5, 2001, pp. B1, B5.

shelf life of smallpox vaccine: Philip K. Russell, "Vaccines in Civilian Defense Against Bioterrorism," *Emerging Infectious Diseases*, vol. 5, no. 4 (July–August 1999), p. 532.

vaccine dilution study: Sue Pleming, "Smallpox Vaccine Studied Due to Terrorism Threat," Reuters, March 14, 2000.

problems with VIG: LeDuc and Becher, "Current Status of Smallpox Vaccine," p. 593; Institute of Medicine, *Chemical and Biological Terrorism*, p. 142; Craig Hooper, "Poxvirus Dilemmas," *New England Journal of Medicine*, vol. 339, no. 27 (December 31, 1998), p. 2027.

West Nile virus: Richard Preston, "West Nile Mystery," *The New Yorker*, October 18–25, 1999, pp. 90–127.

national intelligence estimate: U.S. National Intelligence Council, *The Global Infectious Disease Threat and Its Implications for the United States*, NIE 99–17D, January 2000 <www.cia.gov/cia/publications/nie/report/nie99–17d.html>.

Berger quote: Samuel Berger, "Building on the Clinton Record," *Foreign Affairs*, vol. 79, no. 6 (November/December 2000), p. 32.

AUTHOR INTERVIEWS

Kenneth Alibek, Hadron Corporation, Annandale, Virginia, January 2, 2001.

Judith R. Bale, Institute of Medicine, National Academy of Sciences, Washington, D.C., May 17, 2000.

Kenneth Bernard, National Security Council, Washington, D.C., and Geneva, Switzerland; May 21, 1999 and July 30, 1999.

Lawrence B. Brilliant, SoftNet, Inc., San Francisco, California, August 20, 1999, and April 7, 2000.

Peter Carrasco, Pan American Health Organization, Washington, D.C., March 22, 2000.

Ciro de Quadros, Pan American Health Organization, Washington, D.C., March 22, 2000.

Gerald L. Epstein, White House Office of Science and Technology Policy, Washington, D.C., March 22, 2000.

Joseph J. Esposito, Centers for Disease Control, Atlanta, Georgia, July 19, 1999.

Michael Fitzgibbon, Department of Defense, Washington, D.C., March 20, 2000.

William H. Foege, Rollins School of Public Health, Emory University, Atlanta, Georgia, April 27, 1999.

Former Vector scientist (anonymous), February 2000.

David R. Franz, Southern Research Institute, Frederick, Maryland, November 5, 1999.

Margaret A. Hamburg, Department of Health and Human Services, Washington, D.C., March 20, 2000.

Elisa D. Harris, National Security Council, Washington, D.C., November 5, 1999.

D. A. Henderson, Johns Hopkins University, Baltimore, Maryland, September 14, 1998; December 14, 1998; May 2, 1999; March 25, 2000; and July 1, 2000.

David L. Heymann, World Health Organization, Geneva, Switzerland, May 25, 1999.

John W. Huggins, USAMRIID, Fort Detrick, Maryland, July 6, 2000.

James M. Hughes, Centers for Disease Control, Atlanta, Georgia, July 19, 1999.

Peter B. Jahrling, USAMRIID, Fort Detrick, Maryland, March 24, 2000.

Lt. Col. Robert P. Kadlec, National Defense University, Washington, D.C., June 29, 2000.

Ali R. Khan, Centers for Disease Control, Atlanta, Georgia, May 30, 2000.

Jeffrey Koplan, Centers for Disease Control, Atlanta, Georgia, July 19, 1999.

Joshua Lederberg, Stanford University, Palo Alto, California, February 11, 2000.

James LeDuc, Centers for Disease Control, Atlanta, Georgia, July 19, 1999.

Peter Lewin, Hospital for Sick Children, Toronto, Canada, March 20, 2000.

Brian W. J. Mahy, Centers for Disease Control, Atlanta, Georgia, July 19, 1999.

Franklin C. Miller, Department of Defense, Washington, D.C., March 20, 2000.

Stephen S. Morse, Columbia University, New York City, November 3, 1999.

Michael T. Osterholm, Infection Control Advisory Network, Inc., Eden Prairie, Minnesota, December 22, 1999.

Sergei Popov, Hadron Corporation, Annandale, Virginia, January 4, 2001.

Lev S. Sandakhchiev, State Research Center of Virology and Biotechnology "Vector," March 26, 2001.

Robert Siegel, Stanford University, Palo Alto, California, September 13, 1999.

Robert J. Tosatto, U. S. Public Health Service, Rockville, Maryland, March 26, 2001.

Venkatesh Varma, Indian Mission to the United Nations, Geneva, Switzerland, May 21, 1999.

Mitchell Wallerstein, MacArthur Foundation, Chicago, Illinois, September 1, 1999.

Kenneth D. Ward, Department of State, Washington, D.C., May 25, 1999.

Andrew Weber, Department of Defense, Washington, D.C., November 13, 2000.

Lt. Col. R. Joel Williams, U.S. Air Force, Navarre, Florida, June 12, 2000.

Alan Zelicoff, Sandia National Laboratories, Albuquerque, New Mexico, May 11, 2000.

INDEX